A Systems Approach to Programmable Controllers

FRED SWAINSTON

Delmar Publishers Inc.®

For information, address Delmar Publishers Inc.,
2 Computer Drive West, Box 15–015
Albany, New York 12212

First published in 1991 by Thomas Nelson Australia

Printed in the United States of America
Published simultaneously in Canada
By Nelson Canada
A Division of The Thomson Corporation

10 9 8 7 6 5 4 3 2 1

Library of Congress Cataloging-in-Publication Data

Swainston, Fred.
          A systems approach to programmable controllers / by Fred
      Swainston
                p.   cm.
      Includes index.
      ISBN 0-8273-4670-0
          1. Programmable controllers   I. Title.
   TJ223.P76S93   1992
   629.8'95--dc20
                                                          90-20829
                                                            CIP

Contents

Chapter 13 Programmable controller installations

Chapter 14 Troubleshooting in PLC systems

Preface

As the manufacturing industry moves into the 1990s, only those who have the resolve to move their thinking along with the times will succeed. Automation in tough economic times can improve production output and reduce costs. It is time for the modern workforce to become technically literate and make use of the modern manufacturing techniques that programmable controllers allow. New and exciting electrical and electronic developments allow the application of devices such as programmable controllers to be integrated into a system which can make work easier, safer, and provide a better quality of life while maintaining quality, efficiency, and a market for the product produced.

This book has been written as a course of study that will introduce the reader to the concepts of programmable controllers and explain how they can be used in a system. The written material for the book has been developed from practical and classroom observations and is the culmination of years of providing assistance to those who wish to use programmable controllers in workplace applications. It has been written to be concise and easy to understand, therefore making it a useful text for managers, graduates, technicians, electricians, and shop floor personnel. It is intended to assist those who have a basic knowledge of electrical and electronic principles and practices. It is an advantage if the reader has some knowledge of the manufacturing or process industry. It is expected that the book could be used as a self contained study guide, but would also suit classroom use.

The subjects in the book have been presented in a logical sequence and key points are reviewed in the test questions at the end of each chapter. An instructor's manual, to be used in conjunction with the book, has been developed, and contains additional study material, a syllabus and answers to some of the questions. This manual will assist both instructor and student in assessing if chapter objectives have been attained and in extending their understanding of programmable controllers.

In addition to programmable control the reader is introduced to such concepts as numerical control, computer integrated manufacture, process control and robotics. The reader of this book could feel confident, after studying its contents, of operating in an industrial environment where automation is being applied, and of using the information obtained as a basis for the acquisition of a thorough understanding of any automated system.

Fred Swainston
May 1991

Acknowledgments

I gratefully acknowledge the assistance given to me by the Advanced Certificate students in developing the material used in the contents of this book. Their questions and the difficulties they encountered as they studied this subject have become the basis for many of the examples used throughout the chapters. I extend thanks to my colleague Colin Murphy who advised and assisted with the technical contents of the book and to Reg Varty for his assistance in proof reading. The drawings drafted by David Coppin show his drafting skills and enhance the written text. I extend appreciation to those who were kind enough to review the book, to the Technical and Further Education teachers who provided valuable technical assistance and to Broadmeadows College of TAFE for the use of college equipment. I extend thanks to the PLC manufacturers for allowing use of their photographs and to the staff at Thomas Nelson Australia, in particular Janet Mau, for their advice and co-operation. Finally I extend thanks to my wife Janet and daughters Lisa-Jane and Samantha, who worked long and tedious hours alongside me and were a constant source of encouragement.

Introduction

This book will provide sufficient information to allow the reader to make use of a programmable controller to solve machine and processes control problems. A systems approach has been adopted because either the programmable controller and associated components are a system and should be considered as such, or the programmable controller is used as a component in a much larger system and therefore must be considered in a system context.

The introduction contains four main topics necessary for first-time uses of programmable controllers to make a safe and quick start in getting to know the programmable controller of their choice. A study of the book, through a sucession of hands-on exercises, practical applications, and case studies, will allow the user to become familiar with those concepts common to all programmable controllers.

Programmable controller safety

It is important that safety be kept foremost in the mind when using programmable controllers in a learning situation. Under no circumstances should a programmable controller which is installed to control a machine or process be used for experimental programming or in a learning situation. The programmable controller which is installed has a specific function which, if altered by accident or otherwise, could result in injury to personnel or damage to the machine or process. Throughout this book and the handbook provided by the manufacturer warnings appear where instructions or practices may be dangerous to personnel or may damage equipment. Each warning must be noted and steps must be taken to ensure that accidents do not occur which involve programmable controllers. Discussions with experienced users of programmable controllers will significantly reduce the possibility of accidents.

To assist in developing a sense of programmable controller safety it is suggested that all programming and other exercises even in a simulated or training environment be carried out as though the controlled machine or process is connected to the programmable controller. Become familiar with the programmable controller by connecting it to lamps and switches to simulate outputs and inputs. Develop and test the program for the application before making final connections to the machine or process.

Every programmable controller program in this book has been tested using a programmable controller in a classroom situation. These programs have been specifically selected to demonstrate a range of applications and are intended to provide the reader with programming practice or to demonstrate programming technique. Where possible, the programming examples have been taken from actual programmable controller applications. As there are many variables which occur in each programmable controller system the author cannot assume responsibility or liability for any application that the reader may consider appropriate for a programmable controller. It is, instead, up to the programmable controller user to distinguish between differences in the use of solid state equipment such as a programmable controller and electromechanical applications. In no event will the author be responsible or liable for direct or indirect damage to equipment or injury to personnel as a result of the use or application of the circuits or other information shown in this book.

Acceptable safety design criteria for programmable controllers should include:

- failsafe design;
- permanently programmed sequences that cannot be readily altered;
- no single component whose failure will lead to a hazardous condition;
- directly hardwired emergency stops;
- induced safe-to-start-up checks.

Programmable controller systems can cause the new user difficulties including:

- difficulty with complex hardware;
- difficulty in making them completly failsafe;
- difficulty in making the program inaccessible to unauthorized personnel;
- their susceptibility to electrical interference.

Programmable controller manufacturer handbooks

Programmable controller handbooks contain details necessary to allow the user to install, program and maintain a programmable controller. They will vary from manufacturer to manufacturer in content and detail, but will always contain:

- installation and wiring details;
- program instructions;
- troubleshooting flowcharts and/or techniques.

The first section the user should be concerned about is the installation and wiring information where details of installation and safety requirements will be explained. This section will show how to connect relevant power supplies to the programmable controller as well as the input/output wiring connections which will allow inputs to be accepted by the programmable controller input circuits, and output signals which will be generated by the programmable controller output circuits to drive output devices. The installation manual will also provide hardware details, such as:

- the function of modules;
- hardware addressing;
- system configurations.

The next section of the manufacturer's manual to be considered is the instruction section, termed the instruction set. In this section, each of the programmable controller instructions associated with the device will be explained in detail. An examination of the instructions set will allow the user to determine if the programmable controller set is satisfactory, and if it has a sufficiently extensive instruction set to be able to control the process or machine. The basic set of instructions available in a programmable controller will be:

- relay logic, or digital input and output;
- timer and counter;
- arithmetic and comparison.

These instructions will be available in almost all programmable controllers from the very simple to the very complex.

In programmable controllers, where a more complex instruction set is provided, instructions such as shiftregister, file, sequencer, jump, proportional integral and dirivative, communications, and complex arithmetic will be available.

The third section of the programmable controller handbook to consider is the troubleshooting section. This section will provide assistance when the programmable controller does not function as expected. It will explain such functions as:

- diagnostic instructions;
- status indicators;
- troubleshooting flow charts.

Construction of a programmable controller input/output simulator

When the new user is familiar with the wiring, hardware, and troubleshooting requirements of the system, the construction of an input/output simulation device is advisable so that programming techniques can be developed prior to the programmable controller being put into service.

The simulator should consist of lamps and seven-segment displays to simulate digital outputs. Switches, pushbuttons, toggles, and thumbwheels will simulate digital inputs. If analog inputs and outputs are to be used, power supplies will provide the required voltages and current for the analog input. Analog outputs can be proved by connecting an appropriate resistor across the output terminals, and using a digital voltmeter to measure the output voltage. A simulator setup will allow the majority of the program applications to be tested prior to the programmable controller being connected to the controlled machine or process. Details of the construction of an input/output simulator are shown in the Instructor's manual.

When the simulator is wired to the programmable controller, each instruction should be programmed according to the manufacturer's handbook and proved so that the result is the same as expected.

Chapter objectives

Chapter 1

After reading this chapter you should be able to:

- name the types of control that led to the development of programmable control;
- draw and name the components of a microprocessor system;
- count and carry out arithmetic using hexadecimal, octal, binary, and decimal numbering systems.

xi

Chapter 2

After reading this chapter you should be able to:
- draw and label the block diagram of a programmable controller;
- state the function of each of the assemblies which make up a programmable controller.

Chapter 3

After reading this chapter you should be able to:
- describe the memory organisation of a programmable controller;
- develop a flow chart for a programmable controller program;
- convert relay ladder diagrams to programmable controller ladder diagrams;
- use relay ladder instructions to develop simple programmable controller programs.

Chapter 4

After reading this chapter you should be able to:
- interrupt the status indicators on a programmable controller;
- describe two modes of operation for a programmable controller;
- describe two peripherals which may be connected to a programmable controller system;
- describe the scan techniques as applied to programmable controllers;
- develop programs using basic logic functions;
- state the function of program jumps;
- state the function of program subroutines.

Chapter 5

After reading this chapter you should be able to:
- state the function of timing and counting in industrial control;
- develop programmable controller programs using counters and timers;
- draw the timing diagrams for timers;
- state the problems that can occur using high speed timers in relation to scan time;
- develop a program using timers as oscillators;
- state applications where timers and counters can be used.

Chapter 6

After reading this chapter you should be able to:
- describe those applications where word manipulation offers advantages over bit manipulation;
- state the function of word input and output devices;
- develop programs using components and basic arithmetic functions.

Chapter 7

After reading this chapter you should be able to:
- state the difference between continuous processes and batch processes;
- state the function and difference between primary and secondary transducers;

- state the function of a process control transmitter;
- draw and label a data aquisition system;
- draw and label a data digital-to-analog convertor;
- state the function of differential amplifiers with respect to data aquisition systems;
- describe the type of control as applied to industry applications.

Chapter 8

After reading this chapter you should be able to:
- describe how editing functions are carried out in a programmable controller program;
- carry out word and file moves in a programmable controller program;
- develop programmable controller programs containing shift register and LIFO functions;
- describe the function of stepladder programs.

Chapter 9

After reading this chapter you should be able to:
- state the functions of files;
- develop programmable controller programs using files;
- develop programmable controller functions using input and output word functions;
- develop programs containing sequences;
- state the advantages and disadvantages of multiplexing inputs and outputs.

Chapter 10

After reading this chapter you should be able to:
- describe the data communication techniques applied to programmable controllers;
- state the function of local area networks as applied to programmable controllers;
- state applications of manufacturing automated protocol.

Chapter 11

After reading this chapter you should be able to:
- state the use of flexible manufacturing systems with respect to computer control in industry;
- describe how computer software can be used to develop programmable controller programs;
- state the disadvantages and advantages of computers in an industrial environment.

Chapter 12

After reading this chapter you should be able to:
- describe the difference between numerical control and programmable control;
- list those applications where numerical control is used;
- state how programmable controllers can be used in robotic applications.

Chapter 13

After reading this chapter you should be able to:

- describe the industrial environment as applied to programmable controllers;
- state how input/output connections can be made to a programmable controller;
- describe the documentation that is required for a programmable controller system;
- state the steps required to commission a programmable controller system.

Chapter 14

After reading this chapter you should be able to:

- state the function of programmable controller alarms and status indicators;
- describe the use of instructions provided in a programmable controller instruction set for troubleshooting purposes;
- identify programmable controller maintenance techniques;
- identify programmable controller system faults.

Programmable controller safety requirements must be described by the student with respect to each chapter's objectives.

Machine control, processes, and PLCs

1.1 Automatic machines and process

Industry commonly uses automatic machines and automated processes for discrete item production, batch process, and continuous process. (A process is the method used to develop or produce the required product.) These automatic machines and processes were developed to mass produce products, control very complex operations, or to accurately operate machines for long periods of time. They replaced much human decision, intervention, and observation.

Machines were originally mechanically controlled, then they were electromechanically controlled, and today they are often controlled by purely electrical or electronic means through programmable controllers or computers. The control of machines or processes can be divided into the following categories:

- electromechanical control;
- hardwired electronic control;
- programmable hardwired electronic control;
- programmable control;
- computer control.

1.2 Electromechanical control systems

Electromechanical or relay control was the first step away from purely mechanical control; it is still extensively used in industry today. Electromechanical control systems comprise field devices such as limit switches, electrically controlled pushbuttons, hydraulic or pneumatic actuators, etc. being connected via field wiring to a centrally located relay or contactor panel. The relay or contactor is an electromechanical device made up of a coil and an armature, which operates a set of contacts. These contacts can switch other circuits on or off thereby providing control logic for the controlled system. The relay panel contains all the necessary relays for the required control function, as well as interconnecting wiring. The incoming field wiring is usually connected to the relays via a screw terminal strip.

The relay system has limited application. It is used where a small and simple system is required and where the particular function of the system is unlikely to be changed. A system using relays only lacks timers and counters, and, if it is large and complex, is costly to wire. A complex relay system also takes up a large amount of space.

The advantages of relay systems are as follows:

- they are very reliable and rarely cause faults;
- maintenance and fault finding are simple.

The disadvantages of relay systems are as follows:

- they take up a large amount of rack space;
- large systems are complex, therefore costly to wire;
- it is very difficult to modify the relay logic without a large amount of rewiring;
- they are costly to document, the drawings being usually done by a draftsperson trained to draft electronic and electrical circuits;
- functions such as counters and timers are difficult to obtain;
- eventually the mechanical components wear out.

Relay control circuits

Relay control schematic circuits using standard electrical symbols in ladder diagram format are the usual method of documentation for electromechanical control systems. It is appropriate to examine the ladder diagram format of wiring documentation here

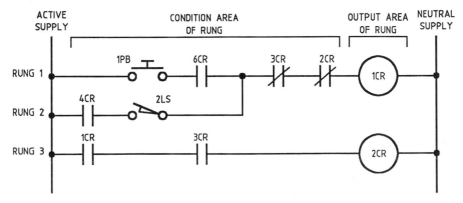

Figure 1.1   Relay ladder diagram

because of its relationship with programmable controller ladder diagram circuits. Figure 1.1 is a diagram of a simple relay control circuit. The circuit is represented by two vertical parallel lines which are the positive and negative of a DC power supply or the active and neutral of an AC supply. The various field devices and control relays are located horizontally, between the vertical lines. Each horizontal connection represents a rung of ladder diagram logic. The field or input devices are on the left of the rung, which is termed the condition area of the diagram, while the right hand side of the rung is where the relays or output devices are located. Each line across the page represents the logic required to operate a relay or similar output device.

The systematic circuit symbols used in the ladder diagram are:

- PB – pressbutton or pushbutton;
- CR – control relay or control relay contact;
- LS – limit switch.

A normally open relay contact is shown by two vertical lines (not to be confused with the electronic symbol of a capacitor). If there is a diagonal line between the two vertical lines on the symbol, the contact is normally closed. Appendix 1 shows commonly used electrical circuit symbols. The numbers and/or letters designating the symbol indicate which control relay is associated with which contact. The numbering system will follow a logical order and include all field and relay devices.

1.3 Hardwired electronic control

As control functions became more complex the need for electronic timers, counters and other electronic devices became apparent to those designing control systems. These electronic devices were designed and made to be mounted in a similar manner to the relays used in electromechanical control and were permanently or hardwired into the control system. Electronic devices are reliable, although system troubleshooting is more complex than in electromechanical control systems. Electronic counters and timers are costly and do not allow wiring flexibility when the function of the system needs to be changed.

1.4 Programmable hardwired electronic control

The introduction of programmable elements such as timers, counters, and sequencers into the control system brought flexibility and multifunction to the system design. Programmable elements allowed the function of the machine or process to be changed, as required, without costly rewiring and documentation. Programmable hardwired control was used prior to the introduction of the programmable controller.

1.5 Programmable logic controller (PLC)

The programmable logic controller (PLC), or programmable controller (PC), is designed to allow field wiring to be terminated on input/output terminals and to be reprogrammed by personnel without special programming skills. Each time the machine function or process is changed, no wiring changes are necessary.

In 1978 the National Electrical Manufacturers Association (NEMA) released a definition of a programmable controller:

a programmable controller is a digitally operating electronic apparatus which uses a programmable memory for the internal storage of instructions for implementing specific functions, such as logic, sequencing, timing, counting and arithmetic, to control through digital or analog input/output, various types of machines or process.

Initially the PLC was used to replace relay logic, but its ever increasing range of functions means it is found in many and more complex applications such as multiple robot systems. The PLC can be used in any industrial application where operating requirements are complex and constantly changing and high reliability is required. It has been designed to be located in a harsh industrial environment.

Programmable controllers are available in a wide range of types. Three types, representing a cross section of the size and options available, are used as examples in this book where programming and specific operations are discussed. The types are:

- Allen-Bradley 2 series PLC;
- Gould Modicon Micro 84;
- Hitachi P 250E.

The programmable controller is an event-driven device, which means that an event taking place in the field will result in an operation or output taking place. A close relative to the programmable controller is a sequence-based controller or sequencer. A programmable controller can be programmed as a sequence controller. (Sequence controllers will be discussed separately in Chapter 9.)

The programmable controller is a special purpose industrial computer which has been designed to be programmed and installed by technical personnel without prior programming knowledge. Programming is easy because it is entered in ladder logic format. Programmable controllers have increased in sophistication and size and many additional functions have been added. Each new generation of programmable controller developed brings the PLC closer in function and programming to the computer.

There are disadvantages in using PLCs. Fault finding is more difficult than with relay logic because PLC systems are often more complex. The fault diagnostic tools available in the PLC system, however, ease this difficulty. Failure of a PLC may completely stop the controlled machine or process, or the PLC memory may be disrupted by external interference. PLCs are, however, inherently very reliable and, once in service and operational, rarely develop faults. Industry has recognized that the high reliability of PLCs does not give maintenance personnel the opportunity to obtain 'hands on' troubleshooting so that when faults do occur they may not have worked on the PLC system for some time. For this reason, ongoing training for maintenance personnel in the PLC system is considered by industry to be essential.

1.6 Computer control

Computers, particularly personal computers, are being accepted as useful devices for machine and process control. Computer costs are comparable with the programmable controller, and computers can handle the complex computations required in some industrial processors. The computer is not, however, designed to operate in a harsh industrial environment, although it can be located in a separate environmentally-controlled area and communicate with devices associated with the machine or process over a transmission medium. Computer-controlled machines are discussed in detail in Chapter 11.

1.7 Microprocessor systems

Microprocessor architecture

Programmable controllers, sequence controllers and computers are based on microprocessors. The commands, instructions and terms used when discussing microprocessors are often those used when discussing programmable controllers and the execution of a program is also similar from one to the other. There are many types of microprocessors. Each has its own unique operational characteristics and all have advantages and disadvantages when compared with one another. Some of the more common microprocessor types are the Zilog Z80, Motorola 6800 series, Intel 8085 series and the Synertek 6502 series. The operation of the microprocessors and the components within their systems are similar. The microprocessor system selected for discussion in this text is the Motorola 6800 series.

Microprocessors are generally discussed in terms of the number of data bits that they can handle simultaneously, i.e. 4, 8, 16 or 32 bits. A bit is a single unit of data. In logic circuits it represents either the 'off' or the 'on' condition, termed a logic 0 or a logic 1. The word 'bit' is derived from the first two letters of 'binary' and the last two letters of 'digit'. The earliest microprocessors were 4-bit devices but advances in technology quickly led to the 8, 16, and 32-bit devices currently available. The greater the number of bits that a processor can handle simultaneously the faster its operation and the more the processor is able to complete complex tasks at high speed. The microprocessor-integrated circuit of a microprocessor system is termed the central processing unit (CPU). A CPU is considered to be the control section of a computing system and contains control functions, instruction encoding, arithmetic unit, registers, accumulators, and control logic. The components inside the microprocessor-integrated circuit are described as the microprocessor architecture. The Motorola 6800 series of microprocessor is an 8-bit device and contains two 8-bit accumulators and an 8-bit code condition register as well as three 16-bit registers (the index register, program counter, and stack pointer (see Figure 1.2)).

To allow information to be accessed (input or taken out of the microprocessor system) the information must have an address or an identification code which separates locations and functions.

Registers and accumulators

Accumulators are temporary storage registers used while the microprocessor is carrying out calculations. They are used for shifting data around the system, and for holding operands (operator commands) and the results obtained during the manipulation of the instructions. The index register is a memory pointer which is used for addressing. The contents of this register are added to the contents at a memory location to produce the actual address of the data. The index register is also used to transfer data from one

5

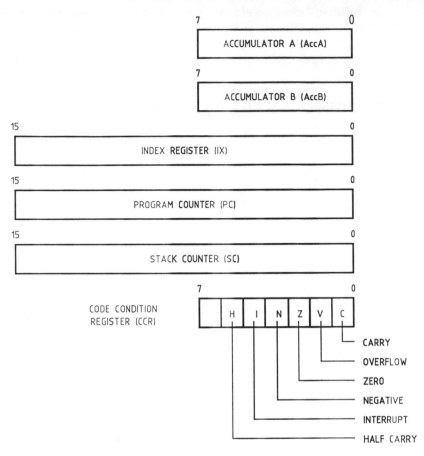

Figure 1.2   Motorola 6800 microprocessor register and accumulator assignment

location to another. The program counter holds the address of the next program instruction to be executed. The CPU, when operating normally, will get the instruction at the address of the program counter and the program will then increment to the next instruction. The program counter contents can be changed so that the next instruction in the program is not the instruction to be operated on next. Such a change in the program execution sequence is termed a program jump. The jump can be to a subroutine (a subprogram accessed by jumping from the main program or routine) which, once completed, causes a jump back to the next instruction in the main program prior to the jump.

The stack is a section of random access memory (RAM) (see Chapter 2, page 27) which may be an external integrated circuit or inside the microprocessor device. It is a last in first out (LIFO) device in which a number of bytes of data can be stacked, and is used to assist the program in handling interrupts and subroutines. (A byte of data consists of 8 bits.) The stack pointer is a register which contains the address of the next vacant location in the stack and will point to where in the program an interrupt or subroutine occurred so that the program location is not lost when returning. (An interrupt is a signal that causes the microprocessor to stop the routine and allows a special routine to take place such as an input from a keyboard.) The code condition register (CCR) flags (signals) when decisions need to be made within the microprocessor system. The flags for the CCR in the Motorola 6800 are as follows:

- Carry – this flag is raised if the last operation produced a carry from the most significant bit; it allows precision arithmetic to be carried out.
- Overflow – this flag indicates that an arithmetic function has exceeded the capacity of a word. A word in the microprocessor consists of 16 bits.
- Zero – this flag indicates if the last operation produced a zero.
- Negative – this flag indicates that the result of the last function was negative.
- Interrupt – this flag indicates that the CPU will allow an interrupt to take place.
- Half-carry – this flag indicates that in the last operation a carry has been generated by the lower 8 bits of the word.

The arithmetic logic unit (ALU) in the Motorola 6800 microprocessor manipulates the data. It performs a wide range of functions such as AND, OR, Exclusive OR, NOT, etc., which are logic functions, as well as arithmetic functions such as add, subtract, increment, decrement, etc. The instruction register holds the operation code (op code) data bits after it has been obtained from the memory so that the correct function is carried out. The output buffer and data buffer circuits are three-state buffers (tri-state) which buffer the data onto and off the address and data buses.

Figure 1.3 shows the block diagram of a microprocessor with all associated address lines, input accumulators and registers. The microprocessor operates by fetching one instruction at a time from the memory, decoding the instruction and then executing the decoded instruction. The instructions are placed or programmed in memory by the programmer as a series of operator codes with an associated address. A simple program for a microprocessor may be an instruction consisting of an operation code in conjunction with an address of the operand. An example of a program where the contents of two memory locations are added is:

Address	Instruction
0000	LOAD X
0001	ADD B
0002	STORE 1
0003	HALT

In this program the contents of memory location X are placed or loaded into the accumulator A. The contents of the B accumulator are then added to accumulator A and the result stored at address 0002. The program is then stopped.

The terms used in a microprocessor program are similar to those used in some programmable controllers. For example, store, load, add, subtract. The program is executed one line at a time from the first to the last instruction unless the program contains a jump, in which case some parts of the program will be missed. The execution of a microprocessor program is similar to the execution of a programmable controller program.

Microprocessor minimal system

A microprocessor does not usually work as a stand alone integrated circuit but requires a number of other devices to be associated with it to make up a microprocessor system.

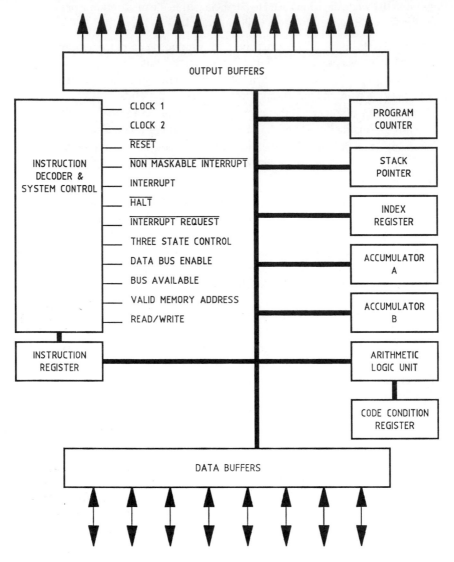

Figure 1.3   Microprocessor block diagram.

A minimal system is one which requires a minimum number of additional components to make it work. Figure 1.4 shows a minimal system based on the Motorola 6800 microprocessor.

The 6800 microprocessor has many applications in process control and data communications. It has seventy-two instructions and six different addressing modes giving it versatility combined with powerful capabilities. The MC 6820 peripheral interface adapter (PIA) easily interfaces with peripheral equipment and the Motorola 6800 series of microprocessor. The interface is achieved through two 8-bit both-way (bi-directional) peripheral interface buses. The clock is used to supply two non-overlapping clock signals required to provide the pulses necessary to make the

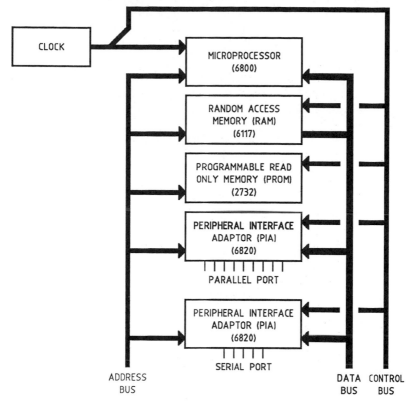

Figure 1.4   Microprocessor minimal system

processor function and provide timing signals for the system. The MM6117 is a static read only memory device (see Chapter 2, page 27) which is used to hold the program written by the user. The 2732 programmable read only memory contains a permanent program which allows the programmer to gain access and operate the microprocessor system. The program stored in this memory will not be lost if power is interrupted to the system.

1.8 Numbering systems

When using programmable controllers, computers or microprocessors, some knowledge of numbering systems other than the decimal numbering system is necessary; the programmer needs to be able to perform conversions between the systems, and to add and subtract within each system. The numbering systems that will be discussed are:

- decimal;
- octal;
- hexadecimal;
- binary;
- binary coded decimal (BCD).

9

The decimal numbering system is base 10, octal is base 8, hexadecimal is base 16, and binary is base 2. Table 1.1 is a comparison between the four numbering systems. Note that all numbering systems start at zero. The system base number is the number in the system where each numbering system produces a more significant digit to the left (see Table 1.1).

In the hexadecimal (base 16) numbering system, letters A to F are used as a convenient way of indicating hexadecimal numbers after 9. At hexadecimal F the system produces a significant digit to the left equal to F + 1 or 10 hexadecimal and continues (see Table 1.1).

Table 1.1   Number system comparisons

Decimal	Octal	Hexadecimal	Binary
0	0	0	0
1	1	1	1
2	2	2	10
3	3	3	11
4	4	4	100
5	5	5	101
6	6	6	110
7	7	7	111
8	10	8	1000
9	11	9	1001
10	12	A	1010
11	13	B	1011
12	14	C	1100
13	15	D	1101
14	16	E	1110
15	17	F	1111
16	20	10	10000
17	21	11	10001
18	22	12	10010
19	23	13	10011
20	24	14	10100

Decimal numbering system

As with all numbering systems, the decimal numbering system (base 10) allows positive and negative (minus) numbers as well as numbers less than one. The first ten whole numbers in the decimal numbering system are 1 2 3 4 5 6 7 8 9 10. In the electronic or electrical field, counting starts at zero (0) and ends at 9, for the first 10 digits. The next number in the decimal system is 1 with a zero at its right indicating 10. At 20 a similar process takes place, the 2 is located to the left of the zero. The left column increments by one every 10 numbers until 100 is reached, a third column is introduced, one column to the left of the previous two digits. The digit furthest left is termed the most significant digit (MSD) and the digit furthest right is the least significant digit (LSD). These rules in the base 10 system are well known and apply to all numbering systems. Figure 1.5 shows the number one thousand, two hundred and thirty-four with the weighting of each digit with the MSD and LSD indicated.

Binary numbering system

The binary numbering system (base 2) is the basis of all digital systems. Two states exist in digital equipment, an 'on' state, which is representative of one (1) and an 'off'

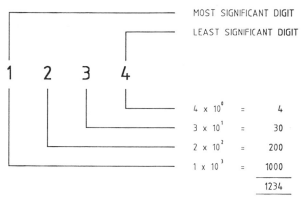

Figure 1.5   Decimal weighting

condition which is representative of zero (0). The on condition in a circuit is approximately equal to supply voltage and the off to zero volts or ground. A third state may exist in some logic circuits to produce tri-state logic. This condition is a high impedance or no voltage state and is not considered in the binary system. The binary numbering system has two numbers, zero (0) and one (1).

Comparing the decimal to the binary system, the decimal numbering system has a radix of 10. This means the numbers to the left of the decimal point are integers (whole numbers) in powers of 10 and those to the right of the decimal point are fractions, in the power of 10. In the binary numbering system the radix is 2 and all numbers are constructed on the basis of the power of 2. To convert binary to decimal, the decimal value of the column of binary numbers are added. The zeros are not added as they have zero decimal equivalent. In the binary numbering system each significant digit has a decimal equivalent of twice the number on its right (see Example 1).

Binary number 421:	1	1	0	1	0	0	1	0	1
Decimal weight of binary:	256	128	64	32	16	8	4	2	1
Decimal equivalent:: 421									

Example 1

To convert a decimal number to its binary equivalent, the decimal number is divided by 2 (the base in the binary numbering system) until the answer is zero. Each time two is divided into the decimal number and a remainder occurs a binary 1 is produced in the binary number; each time two can be divided into the decimal number evenly, a binary 0 is produced. The binary number is read from least significant binary number to most significant binary number, as shown in Example 2.

2) 67 + 1 remainder   LSD
2) 33 + 1 remainder
2) 16 + 0 remainder
2) 8 + 0 remainder
2) 4 + 0 remainder
2) 2 + 0 remainder
2) 1 + 1 remainder   MSD

The binary equivalent of decimal 67 is 1000011

Example 2

Binary addition and subtraction

The same arithmetic principle applies to binary addition and subtraction as for the decimal system. The binary system consists of 0 and 1; any other number generated by the arithmetic will result in a carry to the next column to the left. In Example 3, working from right to left through columns one to six, 0 plus 1 equals 1 for column 1, 0 plus 0 equals 0 for column 2. For column 3, 1 plus 1 is 2 so a 0 is placed in the answer and 1 is carried and then added into the next column to the left. The three ones in column 4 cause 1 to be added to column 5 and 1 to be placed in the answer. Column 5 now has two 1s which produces a 0 in the answer and a 1 is placed in column 6. The subtraction of binary numbers follows the same rules as decimal subtraction using a base 2 (see Example 4).

			Binary					
column	6	5	4	3	2	1		*Decimal equivalent*
		1	1	1	0	1		29
+			1	1	0	0		+ 12
	1	0	1	0	0	1		41

Example 3

			Binary				*Decimal equivalent*
	1	1	1	0	1	0	58
−	1	1	1	0	1		− 29
	0	1	1	1	0	1	29

Example 4

Octal numbering system

The octal numbering system, a base 8 system, is often used in microprocessor, computer and programmable controller systems because 8 data bits make up a byte of information which can be addressed by the PLC user or programmer. In some instances, programmable controller manufacturers use the octal system to number wiring terminals, programmable controller racks, and other PLC hardware. The radix of the octal system is 8, the first eight digits in the system being 0 to 7, so decimal 8 is equal to octal 10. To convert from decimal to octal the decimal number is divided by the system base of 8, as shown in Example 5.

8)196

24   5 octal equivalent of decimal 0.5 or half is 0.4
3   0
0   3

Therefore, the octal equivalent of decimal 196 is 304. Note that when 8 is divided into 196, a decimal remainder of 0.5 results. The decimal remainder must be converted to its octal equivalent by multipling the remainder by the octal base, $0.5 \times 8 = 4$

Example 5

To convert an octal number to its decimal equivalent, each octal digit is multiplied by its octal weighting and then the three values are added (see Example 6).

12

To convert 311 octal to its decimal equivalent:

311

$3 \times 8^2 + 1 \times 8^1 + 1 \times 8^0$

$192 + 8 + 1 = 201$ decimal

The decimal equivalent of octal 311 is 201

Example 6

Octal addition and subtraction

In Example 7, in the right column, octal 6 plus octal 4 equals octal 12. A 2 is placed in the answer and 1 is added to the left column. The addition of the left column is equal to octal 5.

Octal	Decimal equivalent
34	28
+ 16	+ 14
52	42

Example 7

Example 8 shows the subtraction of two octal numbers. Octal 6 cannot be subtracted directly from octal 4, therefore an octal 1×8^1 must be borrowed from the left column. Octal 6 is then subtracted from octal 8 which is equal to octal 2, plus octal 4 equals octal 6 as shown in the right column. An octal 1×8^1 must be returned to the left column to be added to the octal 1×8^1 of the octal 16 so that octal 2 from octal 3 equals octal 1 as shown in the left column.

Octal	Decimal equivalent
34	28
− 16	− 14
16	14

Example 8

Octal-to-binary conversions

As the largest number that can be expressed in the octal numbering system is seven in any given column, octal digits can be represented by three binary bits, i.e. binary coded octal. To convert from octal to binary, the binary equivalent of the octal number is represented by ones and zeros as shown in Example 9.

octal:		7			2			5	
binary coded octal (BCO) weighting:	4	2	1	4	2	1	4	2	1
binary:	1	1	1	0	1	0	1	0	1
decimal weighting of binary:	256	128	64	32	16	8	4	2	1

Example 9

Hexadecimal numbering system

The hexadecimal numbering system is used in programmable controllers, microprocessors and computers because a word of data consists of 16 data bits or two

8-bit bytes. The hexadecimal system is a base 16 system, with A to F used to represent decimal numbers 10 to 15. The techniques used when converting hexadecimal to decimal and decimal to hexadecimal are the same as those used for binary and octal. To convert from a base 10 decimal number to a hexadecimal number the decimal number is divided by the system base of 16 as shown in Example 10.

To convert decimal 28 to its hexadecimal equivalent:

16 $\overline{)28}$

     1 and 12 remainder

The hexadecimal value of decimal 12 is C, therefore decimal 28 is equal to hexadecimal 1C. The remainder left when carrying out these conversions must be multiplied by base 16 to obtain a whole number required for the converted answer.

Example 10

    To convert a hexadecimal number to its decimal equivalent, the hexadecimal digits in the columns are multiplied by the base 16 weight, depending on digit significance. Example 11 shows how to convert hexadecimal number AFC to its decimal equivalent.

$$
\begin{array}{ccc}
A & F & C \\
16^2 & 16^1 & 16^0 \\
2560 & 240 & 12
\end{array}
\quad \text{power of 16 column value}
$$

$$
\begin{aligned}
A = 16 \times 16 \times 10 &= 2560 \\
F = 16 \times 15 &= 240 \\
C = 12 &= \underline{12} + \\
& 2812 \text{ decimal}
\end{aligned}
$$

The decimal value of hexadecimal value AFC is equal to 2812.

Example 11

Examples 12 and 13 show how hexadecimal numbers are added and subtracted.

    To convert a hexadecimal number to its binary equivalent, four data bits are used to represent each hexadecimal number, as shown in Example 14.

Hexadecimal	*Decimal equivalent*
1F	31
+ 92	+ 146
B1	177

Example 12

Hexadecimal	*Decimal equivalent*
FA	250
− 2B	− 43
CF	207

Example 13

Hexadecimal:	A	F	C
Binary:	1010	1111	1100

Example 14

Binary coded decimal (BCD)

The BCD code employs four binary bits with the weights of 1, 2, 4, and 8, to represent each numeral in the decimal system. The four bits on the left of the BCD number represent the most significant digit, whereas the four bits on the right represent the least significant digit as shown in Example 15. Note that the binary is broken into groups of four bits, each group representing a decimal equivalent, therefore the 16 binary bits represent 1001011101000101 = 9745.

Most significant digit   *Least significant digit*

| 1001 | 0111 | 0100 | 0101 | BCD |
| 9 | 7 | 4 | 5 | Decimal equivalent |

Example 15

1.9 Test questions

Multiple choice

1 A process system is controlled without programmable elements and has discrete timers and counters. This system would be considered to be:
   (a) electromechanical control.
   (b) hardwired electronic control.
   (c) programmable control.
   (d) computer control.

2 The device for connecting inputs and outputs to a microprocessor is:
   (a) a CPU.
   (b) a PIA.
   (c) a RAM.
   (d) an ALU.

3 A change in the numerical sequence of program instructions is termed a program:
   (a) fault.
   (b) jump.
   (c) change.
   (d) fetch.

4 Which of the following would be found in a microprocessor?
   (a) Interrupts.
   (b) Accumulators.
   (c) Register.
   (d) PROMs.

5 What is the decimal equivalent of hexadecimal FC?
   (a) 111
   (b) 132
   (c) 207
   (d) 252

6 What is the numerical value of the BCD 001101110110?
   (a) 376
   (b) 673
   (c) 794
   (d) 883

7 What is the decimal value of binary 01101110011?
   (a) 376
   (b) 673
   (c) 794
   (d) 883

8 What is the radix of the decimal system?
   (a) 2
   (b) 8
   (c) 10
   (d) 16

9 What is the octal value of hexadecimal F2?
   (a) 242
   (b) 194
   (c) 308
   (d) 362

10 What is the binary value of hexadecimal F2?
   (a) 11110010
   (b) 11111010
   (c) 11001111
   (d) 0101111

11 The stack is an area of random access memory which handles data as:
   (a) an accumulator.
   (b) an arithmetic logic unit.
   (c) a first in first out.
   (d) a last in first out.

12 What is the decimal equivalent of octal 103?
   (a) 76
   (b) 83
   (c) 94
   (d) 67

Written response

13 List two advantages of a programmable controlled system over a hardwired relay controlled system.

14 What is the major difference between the programmable controller used in an industrial application and a personal computer?

15 Within the code condition register (CCR) there are flags. What is the purpose of a flag?

Programmable controller hardware

2.1 Programmable controller block diagram

Programmable controllers are designed for ease of installation, wiring, and programming. The range of PLCs available is large, with a type to suit all process industrial applications. There are six components found in all programmable controller systems:

- processor;
- input assembly;
- output assembly;
- power supply;
- rack or mounting assembly;
- programming unit.

Figure 2.1, the Allen-Bradley programmable controller system, shows the six components of programmable controller; Figure 2.2 shows a block diagram of the PLC system and the interconnection of each of the six components.

Figure 2.1   Allen-Bradley programmable controller

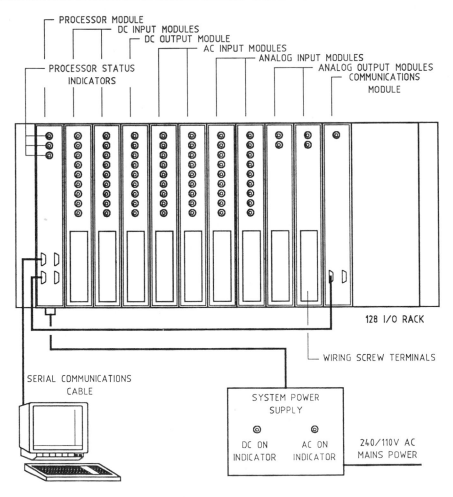

Figure 2.2   Block diagram of PLC system

2.2 Input/output assemblies

There are many types of inputs and outputs that can be connected to a programmable controller, and they can all be divided into two groups: digital and analog. Digital inputs and outputs are those that operate due to change in a discrete state or level. Analog inputs and outputs change continuously over a variable range. Some programmable controllers have separate modules for inputs and outputs; others have the inputs and outputs connected as an integral part of the controller. Digital inputs accept contact operations from the field such as limit, float, pressure and thumbwheel switches. Digital outputs operate lamps, indicators, relays, solenoids, motors, etc. Programmable controllers also have a number of specialist inputs and outputs or modules that allow analog values to be fed into or fed out of the PLC. Analogs as inputs and outputs are discussed in Chapter 7 with reference to process control and data acquisition. The field wiring to the programmable controller system is connected to the digital or analog input/output assembly terminals.

Input assemblies

There are two forms of input signal, AC and DC. Input assemblies are required to accept high voltage AC or DC, and low voltage DC signals. The input assembly has a number of common sections (see Figure 2.3) the functions of which are as follows:

- The input indicator shows the on/off status of the input to the assembly.
- The electrical isolation provides protection to the PLC internal logic by isolating it from the field wiring and terminations.
- The input filter provides field contact debouncing which will assist in reducing multiple unwanted field contact closures due to the characteristics of mechanical contacts. In addition, the input filter will reduce the possibility of electrical noise being detected as a field contact operation.
- The logic circuits process the input signal to make it suitable to be fed into the PLC processor.

The AC input assembly, shown in Figure 2.4, has a neon lamp at the input to indicate it is in an active state. A full wave rectifier bridge converts the AC input signal to DC to operate the optical isolator. Diode D5 is a reverse polarity protection diode which will stop reverse voltages being fed into the isolating device. The isolating device in this application is an optical type. It provides at least 1500 volts of isolation between

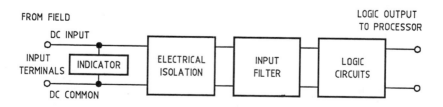

Figure 2.3   Input assembly block diagram

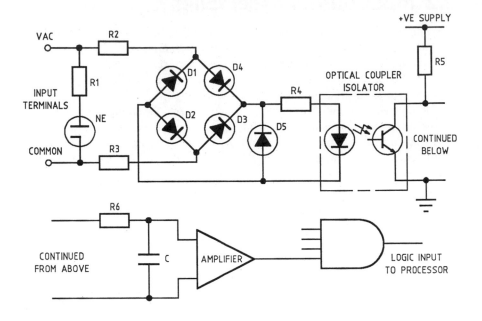

Figure 2.4   AC input

the input terminals and the logic circuits within the programmable controller. The optical coupler has an internal light-emitting diode which, when active, causes a light-sensitive transistor to produce an output. The optical coupler is ideal in this application as there is no electrical connection between the input of the coupler and its output. Where isolation is provided by a transformer, the signal is transferred from the input to the PLC logic via the primary and secondary of the electrically isolated transformer. This type of isolation is termed magnetic or electromagnetic induction isolation and effectively means that there is no physical electrical connection between the field wiring and the PLC processor logic. The action of R6 and C in conjunction with the amplifier provides contact debouncing and also acts as a low pass filter to make the input less sensitive to electrical noise. The output of the amplifier is fed into a logic device suitable for converting the signal to logic levels necessary for processor operation.

The DC input assembly shown in Figure 2.5 is similar in construction and operation to the AC input assembly with the following exceptions:

- No bridge rectifier circuit is required as the input signal is DC.
- A light emitting diode (LED) is used as an input status indicator.

In Figure 2.5, resistor R2 limits the input current. Excessive current through the light-emitting diode will damage it or cause it to fail. D2 is the reverse polarity protection diode. Capacitor R4 and C provide contact debounce and low pass filtering to reduce electrical noise. The field devices are connected to the input terminals. The power supply to provide the required field input voltage is usually a separate power supply provided by the programmable controller user (see Figure 2.6).

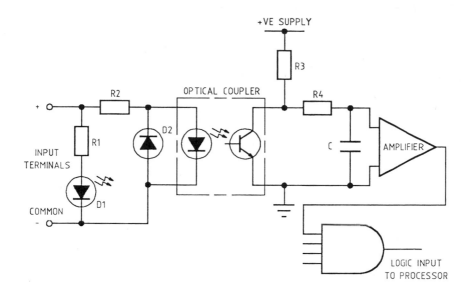

Figure 2.5   DC input

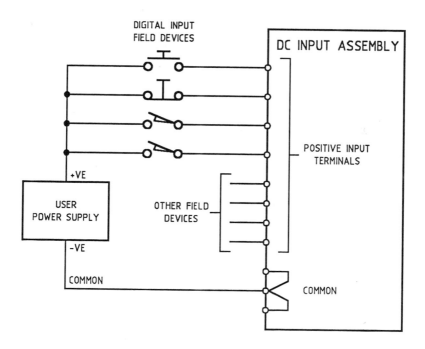

Figure 2.6   Input connections

21

Output assemblies

The output assembly of a programmable controller consists of a number of sections (see Figure 2.7). The output signal is derived from signals in the processor being fed to the output logic circuits which are isolated to protect the logic from accidental application of excessive voltages to field wiring. The output is usually provided with over current protection such as a fuse to stop the output assembly from being damaged if an accidental short circuit is applied in the field wiring.

A blown-fuse indicator is provided as a feature on some programmable controllers; it will indicate if any of the individual output fuses blow and assist in troubleshooting by indicating the general location, such as rack or module, of a fault.

The output assembly of a programmable controller acts as a switch to supply power from the user power supply to operate the output. The output, under the control of the program, is fed from the processor to a logic circuit that will receive and store the processor command that is required to make an output become active. It is necessary to store the command because once an output is made active it must remain in the active state until it is instructed by the program to become inactive. The output switching devices most often used to switch power to the load in programmable controllers are:

- relay for AC or DC loads;
- triac for AC loads only;
- transistors for DC loads only.

The output assembly with a relay output provides contacts to switch power to the load. The contacts are not referenced to earth or supply, therefore providing complete electrical isolation between the programmable controller and the load. The disadvantages of relay contact and in particular reed relay outputs are considered to be:

- The amount of current that can be passed through the relay contacts is limited.

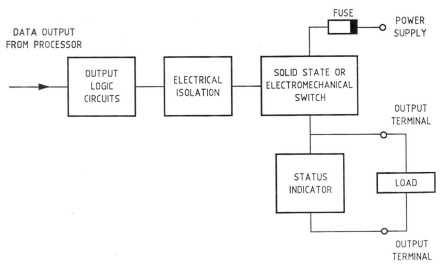

Figure 2.7   Output assembly block diagram

- The time taken for the relay to operate is significant.
- There is electrical interference and relay contact due to sparks being produced as the relay contacts open and close.

To overcome the problems caused by the relay outputs, solid state output devices for AC and DC have been extensively used in programmable controllers. The triac is a three-terminal semiconductor device that is used to switch AC outputs of relatively high current at high voltages. A typical triac will switch 240 volts at 5 amps with ease, thus making it ideal for this task. The symbol of the triac is shown in Figure 2.8(a). The action of the triac is that of a bidirectional switch. When a positive or negative trigger voltage of approximately 2.5 volts at 5 mA is applied to the triac gate, the triac will conduct fully in both directions and will continue to conduct while the voltage is applied to the gate. When the gate voltage is removed, as the next half cycle of the switched AC voltage approaches zero crossover, the triac ceases to conduct. The voltage to be applied to the gate of the triac can be developed from an optical coupling device.

Transistors are solid state devices used to switch the DC outputs in programmable controllers. The type of transistor used is NPN silicon, designed for general purpose switching applications. These transistors would typically be able to switch 30 volts at 2 amps. To cause the transistor to conduct, a bias voltage positive at the base with respect to the emitter is applied. The base/emitter bias voltage will cause the transistor to turn on, thus allowing load current to flow in the emitter/collector circuit. When the base/emitter bias voltage is removed, the transistor will turn off. The symbol of an NPN transistor is shown in Figure 2.8(b). The disadvantages of solid state switching are that outputs are connected to a single common supply which does not allow complete isolation between outputs of the switched circuits, and that a short circuit across the PLC outputs could damage the semiconductor device.

Figure 2.9 shows the output circuits with each of the three types of switching. Note that the relay relies on its contacts for isolation and has no optical isolator, whereas solid state switching employs optical isolation.

A   B

Figure 2.8(a)   Triac symbol   (b)   Transistor symbol

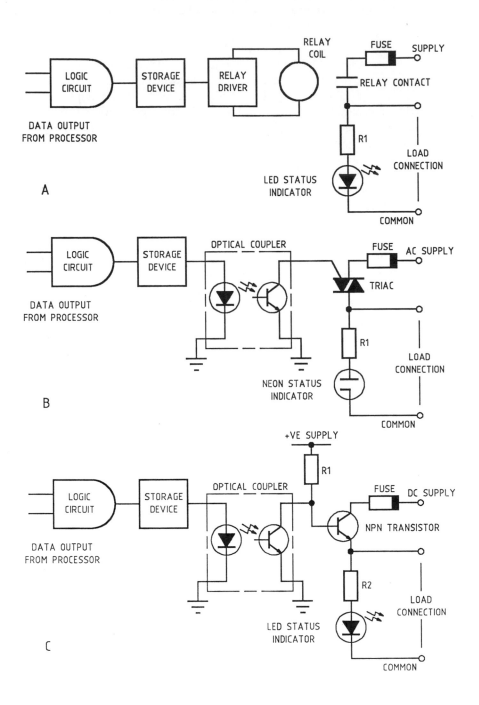

Figure 2.9(a)   Relay output   (b)   Triac output   (c)   Transistor output

Most programmable controllers can be fitted with input and output assemblies compatible with transistor transistor logic (TTL) devices. This allows the programmable controller to directly interface with a wide range of TTL logic devices.

2.3 Processor unit

The processor unit contains a microprocessor system which contains the system memory and which is the PLC system decision-making unit. The processor unit examines by scanning data from the input and output modules and stores their conditions in the memory. The processor unit then scans the user program stored in the memory and at the same time makes decisions to accept input changes and cause outputs to change during the next input/output scan. The scanning process is a cyclic procedure and, unless interrupted by an external command, the processor will continue to scan through the program continually updating inputs and outputs under the control of the user program. The time taken to complete one scan is the total of the time taken to scan the inputs, outputs, and user program. It is the processor unit that performs the counting, timing, data comparisons, sequential operations, and other functions provided by programmable controllers.

Processor operational indicators

Associated with the processor unit will be a number of switches which change the processor's modes of operation to allow the operator to program, run or test the programmable controller system. The two basic modes of operation of a programmable controller are the run mode and the program mode. In the program mode, the processor will allow change to the user program but the programmable controller will not operate, therefore input changes will not produce output changes. Changes undertaken in this mode are termed 'off line'. In the run mode, changes to the program will not be allowed because the user program will be running. A third mode of operation which allows the user program to be tested is included in some programmable controllers. The test mode allows the user to test run a program without operating the outputs. Program changes can not be made when in the test mode.

An 'on line' programming feature is provided by some programmable controller manufacturers which allows programming changes to take place while the processor is executing the user program. If this feature is provided on a programmable controller, extreme care must be exercised by anyone carrying out on line programming changes; changes to the user program while the PLC is operating could injure personnel and damage equipment. In many instances programmable controller manufacturers provide a key lock switch so that only authorized personnel can make programming changes. This mode of operation is termed the run/program mode.

A remote programming facility is provided with PLC systems that makes use of individual programmable controllers spread throughout a large working area. This facility allows the PLC to be programmed from a distant location, such as a control room, over a data communications network (see Chapter 10).

The processor unit will have a number of status indicators to provide system diagnostic information to the operator. The indicators available are:

• processor scan – which indicates that the user program cannot be scanned by the processor;

- memory fault – which indicates that the memory data is corrupt or cannot be transferred through the programmable controller system correctly;
- run indicator – which indicates that the processor is carrying out the program correctly while it is switched to the run mode.

Memory

The memory that stores the user program, as well as the operating program, may be one of a number of types that are available. The method employed to hold programs which can be removed from the computer or programmable controller is called mass storage. The common forms of mass storage in use are magnetic tape, floppy disk, hard disk, and laser disk. Magnetic tape stores programs in a similar manner to the way audio information is stored on magnetic tape. The program is encoded by the computer into a number of audio tones and recorded onto the magnetic material. Once stored, the information can be retrieved at any time by playing the tape back into the PLC which decodes the audio tones.

Floppy disks, sometimes called diskettes, are available in 8-inch, $5^1/_4$-inch, and $3^1/_2$-inch diameter types. The floppy disk is coated with material similar to magnetic tape and has a similar storage and retrieval scheme except that the tracks are located in rings around the disk. The floppy disk is much faster and has much greater storage capacity than magnetic tape, although the disk reader and associated circuitry is complex and more expensive than a magnetic tape recorder. The major disadvantage of floppy and magnetic tape storage is that the stored information can be corrupted by contaminants, such as dirt or grease, and by exposure to magnetic and electric fields. All these contaminants are, of course, common in industrial environments and need to be well-controlled when tapes and floppy disks are stored.

The hard disk is usually associated with computers and not programmable controllers. The hard disk is similar in construction to the floppy disk but it has much larger storage capability and is not usually physically removed from the computer.

The laser disk can store very a large amount of information on a compact disk (CD). The information is embedded into the surface of the disk and is not susceptible to dirt, grease, electric fields or magnetic fields. The disk is read using a laser beam to distinguish the information on the surface of the disk. The laser disk has the disadvantage that once the program is created it cannot be changed without replacing the entire disk. The laser disk is also known as a compact disk read only memory (CDROM).

Core, magnetic bubble, and random access memory are the memory types which usually hold the PLC user program. They will be either permanent or temporary (sometimes termed volatile). Permanent memory will retain the information in the memory when power is removed, whereas temporary memory will lose the information when the power to the memory is removed unless some form of back-up memory power supply is provided.

Core memory

Core memory is a magnetic memory comprising a number of ferrite inductors which will represent a binary 1 if the magnetic field is in one direction and a binary 0 if the magnetic field is in the opposite polarity. (The binary 1s and 0s are the bits of information which make up the program.) Wires are connected through the center of the core to allow reversal of the magnetic field. The memory contents will be retained when power to the core is removed. Core memory is low cost, but it has a slow operating speed compared with other types of memory.

Magnetic bubble memory

Magnetic bubble memory (MBM) is a nonvolatile type of solid state memory consisting of a cylindrical magnetic domain termed a bubble. The absence of a bubble in a specific position represents logical 0 whereas presence of the bubble represents logical 1. Changing magnetic fields are used to move the bubble within the magnetic material, similar to a cyclic shift register. Bubble memory dissipates very low power, typically 1 microwatt per bit, and is faster and considered more reliable than the floppy disk system because it has no moving parts.

Random access memory

Random access memory (RAM) is a type of solid state memory contained in an integrated circuit. The size of the memory is measured in the number of kilobits of information that the memory can hold. A single integrated circuit may contain 1024 bits of information (1 k of memory), two integrated circuits makes 2 k of memory, and so on. The memory is used to store variable information in the form of a program. The microprocessor, under the control of the program, allows input to the RAM which can change the stored information. Random access memory is considered volatile, that is, the information in the memory will be lost if the power to the memory is interrupted.

There are two types of RAM: static and dynamic. Static RAM stores the program in the form of 1s and 0s in memory cells similar to conventional bistable flip flops. Provided power is connected to the static RAM integrated circuit, the memory is maintained. A memory back-up battery is used to maintain the information in the memory for the duration of power failures. The power supply current to operate static RAM is more than the power supply current required for dynamic RAM operation.

Dynamic RAM stores the program in 1s and 0s as electrical charges on the gate capacitance of metal oxide semiconductor (MOS) transistors. The charge on the gate capacitance will gradually leak away due to the capacitors not being perfect, so dynamic RAM must be electronically refreshed. The refresh circuitry makes dynamic RAM operation more complex and difficult to synchronize compared to static RAM.

Read only memory

Read only memory (ROM) is a nonvolatile memory and is used to store permanent housekeeping programs in the programmable controller. The housekeeping program is the program that makes the PLC function. It consists of tables, function generators, diagnostic programs, error messages, and generally the programs that make the PLC user friendly. Read only memory describes a wide range of devices in which 1s and 0s are located in the form of a program. The ROM will not lose the stored program even if it is removed from the printed circuit board. It has a matrix of memory cells each of which are addressable. The types of ROM are:

- mask programmed;
- programmable read only memory (PROM);
- erasable programmable read only memory (EPROM);

- electrically alterable programmable read only memory (EAPROM);
- electrically erasable programmable read only memory (EEPROM).

ROM is much easier to interface with microprocessor systems than RAM as most ROM are static devices, need no clock, and operate from a single power supply.

Mask programmed ROM are programmed during manufacture to user requirements or specifications. The program bit pattern, once determined, is converted to a mask which is permanently installed into the ROM integrated circuit during the final layer of metalization while the ROM integrated circuit is being manufactured. Once the ROM is made it cannot be changed. The cost of this type of ROM is high and it is only used when large numbers of the same ROM are required in the system.

Programmable read only memory (PROM) stores its program as 1s and 0s in memory cells. The program is inserted into the PROM by a selective fusing technique which opens circuit links within the PROM using an electronic PROM programmer. When the PROM is programmed the program is permanent and cannot be altered.

Erasable programmable read only memory (EPROM) has an advantage over the previously discussed ROM in that it is erasable. The user can erase the program in the EPROM by removing the EPROM from the printed circuit board and exposing it to ultra violet light for 10 to 20 minutes. The top of the EPROM integrated circuit has a window through which the ultraviolet light can pass. The program in the EPROM is stored as a charge in a metal oxide semiconductor field effect transistor (MOSFET) memory cell. PROMs that can have the program stored within them changed by a special process, such as an EPROM, are also termed read mainly memory (RMM).

Electrically alterable programmable read only memory (EAPROM) can be reprogrammed by applying voltages to the memory device, using circuit techniques not normally found in conventional memory circuitry, without removing the EAPROM from the printed circuit board. It takes some milliseconds to alter the program, much less than the time taken to change an EPROM. Programming the EAPROM is similar to the EPROM.

Electrically erasable programmable read only memory (EEPROM) is programmed in a similar way to the EPROM, but the EEPROM can be erased by applying special voltages to memory integrated circuits while it is still located in the printed circuit board. The time taken to erase this type of memory is much shorter than exposing conventional EPROM to ultraviolet light.

Memory size

The programmable controller program is stored in the memory as 1s and 0s which are assembled in the form of 16-bit or 8-bit words in an 8-bit microprocessor system. The memory size of a programmable controller varies from as small as 256 words for small systems to 128 000 words for a very large system. Memory sizes are expressed in thousands of words that can be stored in the system; thus 2 k is a memory of 2000 words, 64 k is a memory of 64 000 words and so on ('k' is the abbreviation for kilo or one thousand). Most programmable controllers have memory expansion modules or socket space available on the processor printed circuit board to expand the memory size if necessary.

2.4 Power supply

The programmable controller system requires two power supplies. One provides power necessary for field devices and output loads to operate and is provided by the programmable controller user. The second power supply is provided internally or as a module which is part of the programmable controller system. This power supply provides internal direct current (DC) to operate the processor logic circuitry and input/output assemblies. The voltage that it provides will depend on the type of integrated circuits (ICs) within the system. If the system is made up of transistor transistor logic (TTL) ICs, the internal power supply will be 5 volts, but if the integrated circuit is a complementary metal oxide semiconductor (CMOS) type, the power supply voltage will be in the range of 3 volts to 18 volts. The power supply will be regulated as well as having over voltage and over current protection. It will be designed to operate from a 110 volt or 240 volt, 60 or 50 Hertz (Hz) alternating power source. The operational input voltage of the power supply is selectable by strapping or switching and can be specified when the system is purchased. The power supply source voltage will have a fuse located at the input to the power supply. Status indicators associated with the power supply may be an input or voltage source indicator which indicates that the AC voltage source is connected and available at the input to the power supply, or a DC voltage indicator to indicate that the power supply is producing the necessary DC voltage for the processor and the input/output assemblies within the system.

If the RAM is volatile a back-up battery is used to supply power to the memory during power failures. The cells may be a NICAD secondary or zinc/carbon primary, grouped so as to provide the required voltage to maintain memory when power failures occur. The battery may be located in the power supply, processor or may be separate from the PLC system. Some programmable controller manufacturers provide a battery-low indicator. Batteries should be checked and/or replaced on a regular basis in line with the manufacturer's maintenance procedures.

2.5 Programming units

The programming unit allows the programmable controller user to enter, edit and monitor programs by connecting into the processor unit and allowing access to the user memory. The programming unit can be a liquid crystal display (LCD) hand-held type to terminal, a single line LED display unit or a keyboard and a visual display unit (VDU). The programming unit communicates with the processor via a serial or parallel data communications link (see Chapter 10). If the programming unit is not in use it may be unplugged and removed. Removing the programming unit will not affect the operation of the user program. Figure 2.10 shows the programming unit for the Allen-Bradley program unit dedicated to the programmable controller. A computer with appropriate software can also act as a program terminal, making it possible to carry out the programming away from the physical location of the programmable controller. When the program is complete it would be saved to some form of mass storage and down loaded to the programmable controller when required. During the time that the computer is not being used for PLC program development it can be used for other purposes. A dumb terminal, via a data communications link, can also be used as a remote programming terminal for the PLC. The dumb terminal has no inbuilt intelligence (software) but works under the control of the PLC processor.

Figure 2.10   Allen Bradley programming unit

2.6 Programmable controller system assemblies

Programmable controller manufacturers have approached the problem of producing a very reliable programmable controller that can operate in a wide variety of industrial environments by using two separate design characteristics. One uses modules for each section as shown in Figure 2.11 in the Modicon Micro E84 system. The second uses a stand alone complete programmable controller unit, for example the Hitachi P 250E shown in Figure 2.12.

Programmable controllers which use a rack to hold a number of modules are usually found in systems where a large number of inputs and outputs, and a variety of types of input and output, such as a mixture of analog and digital modules, are required. Each module is plugged into an edge connector which provides intermodule wiring via a mother board or cable at the back of the rack assembly. The cable connections to inputs and outputs are via screw terminals on the front of each module.

The nonmodule type of programmable controller has interconnections between modules internally hardwired and has screw terminals for field wiring. It usually has fewer input and output connections and has less variety in the types of input and output assemblies compared to the module type.

It is a safety requirement that the overall programmable controller assembly be earthed according to local mains authority specifications because the power supplies operating the system will be 110 volt or 240 volt AC.

Figure 2.11   Gould Modicon Micro 84 PLC

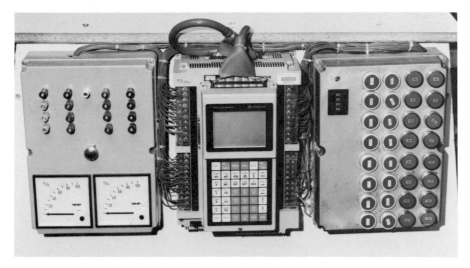

Figure 2.12   Hitachi P250E PLC and input/output simulator

2.7 Test questions

Multiple choice

1 Which of the following are considered to be disadvantages of relay contact outputs used in some programmable controllers?
   (a) No earth reference at the output.
   (b) The time it takes the contacts to operate.
   (c) The electrical interference produced.
   (d) The amounts of current that can be passed through the contacts.

2 The voltage applied to the gate of a triac:
   (a) allows it to conduct.
   (b) stops it conducting.
   (c) allows positive AC cycles to pass.
   (d) stops positive AC cycles from passing.

3 Which of the following types of memory are considered to be volatile?
   (a) Core memory.
   (b) Random access memory.
   (c) Read only memory.
   (d) Bubble memory.

4 Which of the following devices can be described as a read mainly memory (RMM)?
   (a) Core memory.
   (b) Random access memory.
   (c) Erasable programmable read only memory.
   (d) Masked programmable read only memory.

5 Which of the following components make up a programmable controller system?
   (a) Input/output assemblies.
   (b) Programming unit.
   (c) Core memory.
   (d) Scan counter.
   (e) Processor.

6 The function of the optical coupler in a programmable input assembly is to:
   (a) keep noise out of the PLC system.
   (b) allow the assemblies to operate using AC or DC.
   (c) convert the input signals to logic levels.
   (d) provide high voltage protection for the PLC system.

7 The function of the rectifier bridge in the AC input module is to:
   (a) provide electrical isolation.
   (b) convert the input signal to logic levels.
   (c) convert the AC input to DC.
   (d) improve noise immunity.

8 The type of output available in programmable controllers are:
(a) bridge rectifier.
(b) relay.
(c) triac.
(d) transistor.
(e) TTL.

9 Status indicators are provided on each output of the output assembly to indicate that:
(a) load has been operated.
(b) output is active.
(c) input associated with the output is active.
(d) assembly fuse has blown.

10 EPROMs can be erased by:
(a) applying special voltages to some IC pins.
(b) exposing them to electrostatic fields.
(c) fusing links inside the EPROM.
(d) exposing them to ultraviolet light.

True or false

11 Read only memory is volatile.
12 EEPROM is electrically enterable programmable read only memory.
13 The power supply current for dynamic RAM is more than the power supply current required for static RAM.
14 Dynamic RAM needs to be refreshed.
15 Triacs are used to switch DC outputs.

Written response

16 Describe the function of the backup battery.
17 Define a bit of information with respect to information storage in a programmable controller.
18 Describe the advantages and disadvantages of static as compared to dynamic RAM in programmable controllers.
19 There are often two power suppliers used in conjunction with PLCs. Write a statement which describes the function of each.
20 List the categories of functions that can be carried out using the PLC programming unit.

Relay ladder diagrams and programming

3.1 Memory organization

To be able to program a programmable controller it is necessary to have some knowledge of how the memory of the programmable controller is organized. When the program is being written the user must know exactly where the program will be located in the PLC memory and how to address inputs and outputs.

The size of the programmable controller memory relates to the amount of user program that can be stored. The memory sizes range from as small as 0.5 k which can hold 512 16-bit words, or a total of 8192 bits of data, to 128 k which can hold 128 000 16-bit words or a total of 2 048 000 bits of data. Memory sizes can be larger, but the programs for programmable controllers generally do not require storage space above 128 k and in many instances only need a memory size of 1 k to 2 k. The memory of the programmable controller can be divided into three areas: storage memory, user memory, and housekeeping memory. The storage memory and user memory make up the programmable controller's data or register table. (The term used to describe the data storage tables depends on the PLC manufacturer.) The storage memory is used to orientate real world input/output devices, counter/timer accumulated and preset values, and bit or word storage for internal relay to a specific memory location. The user program area of memory is set aside and used to store the user program which causes the programmable controller to operate in a particular manner. The user program is stored in a number of 16-bit words, each word containing one program instruction. The bits within an instruction word will be 'on' or 'off', thus providing each word with program instruction meaning. The housekeeping memory is an area of memory that cannot be accessed by the user. It is used during the programmable controller operation to carry out functions which are necessary to make the processor operate, to carry out arithmetic functions, and to carry out other internal operations. Figure 3.1 shows the general organization of a programmable controller memory.

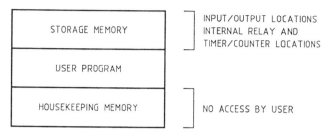

Figure 3.1   General organization of programmable controller memory

Programmable controller word organization

The use of storage bits or words in programming is considered to be one of the major advantages of programmable controllers as it allows the development of complex relay logic without the cost and wiring expense of a physical relay. Programmable controller words are made up of 16 bits of information as well as several bits which are used for internal checking of the processor system to ensure accuracy and a means of detecting errors. The additional bits are often termed redundant bits or parity bits in the programmable controller word. A bit within the word can only exist in two states, a logical 1 or 'on' condition, or a logical 0 or 'off' condition.

Figure 3.2 shows the word structure of an Allen-Bradley PLC word. It consists of 16 data bits and 2 parity bits. The first 8 bits of the word from 00 to 07 are the lower byte, and bits 10 to 17 are the upper byte. Each word is identified in the programmable controller memory by a unique word address and each bit by its bit number, therefore allowing any single bit within the accessible memory to be addressed. The words of a programmable controller memory organization table are directly controlled and manipulated by the processor.

Figure 3.3 represents the factory configured data table for the Allen-Bradley PLC 2 series of programmable controllers. Two processor work areas are provided for housekeeping functions. The output image table is the location in memory where out-

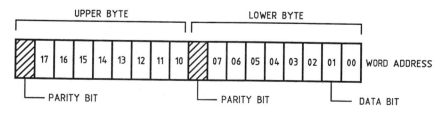

Figure 3.2   Word structure

35

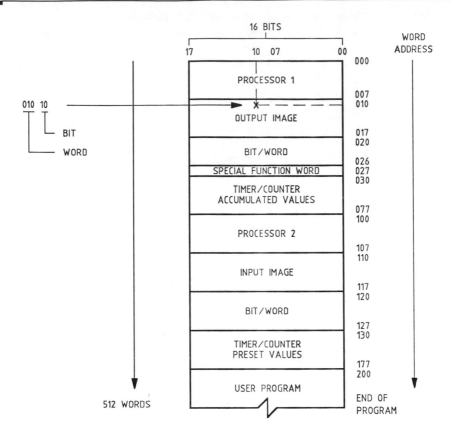

Figure 3.3   Allen-Bradley memory organization

puts are identified. An output can be located using a cross reference between the three-digit octal word and two-digit octal bit address. Under the control of the user program a bit in the output image table can be turned 'on' or 'off' to perform an operation of a field device wired to the output via the output module. If there are no physical outputs wired to a bit address, the address can be used as an internal storage point. Internal storage bits or points can be programmed by the user to perform relay type functions, therefore providing the logic necessary to control an operation. The advantage of an internal storage bit is that it does not unnecessarily occupy a physical output.

The input image table transfers the field input status to a location in the memory via the input module. The status from the field digital inputs will be either 'on' or 'off'. The status bits of the input image table are monitored by the processor, and, as a change in input status occurs, the processor under the control of the user program will take the appropriate action.

Two memory areas are used for bit or word storage and two for counter and timer programming, one containing the accumulated value of the counter or timer and the other containing the preset value. Counters and timers will be discussed in Chapter 5.

A message storage area and subroutine area are located at the end of the user program in the data table organization. The message storage area is used to contain messages which relate to various parts of operating programs and can be entered as required by the user. When the user program is too large and complex, messages which include program information about the function of the program are often used to make the program operation easier for the user to understand. If the message area is not used, it can be reduced in size by reconfiguring the memory. Not all programmable controller memories can be reconfigured.

The subroutine areas allow the programmer to program a jump from the main program to a subroutine program and at the end of the subroutine jump back to the next line in the main program. Subroutine programs are sections of program that are accessed on an 'as required' basis. They are discussed in Chapter 8.

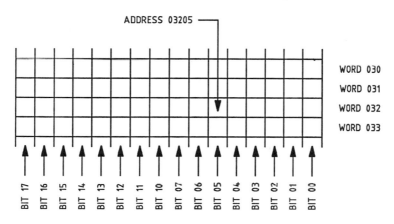

Figure 3.4   Allen-Bradley word and bit distribution

The Allen-Bradley memory, as it is when factory configured, has space for 512 16-bit words, or can contain 8192 bits of information. Each word has a three-digit octal address. The three-digit octal address, in conjunction with a two-digit octal bit address, locates any bit in the data table (see Figure 3.4). In the Allen-Bradley system, a five-digit octal code locates any input or output on the data table. For example, the address 03205 will appear on the data table at word location 032 and bit location 05. To allow ease of programming Allen-Bradley have divided the address so as to represent various areas of the programmable controller, for example, address 01402 indicates as follows:

0 – output
1 – rack 1
4 – module group 4
0 – output terminal number most significant digit
2 – output terminal number least significant digit

An input address 11215 would be as follows:

1 – input
1 – rack 1
2 – module group 2
1 – input terminal number most significant digit
5 – input terminal number least significant digit

37

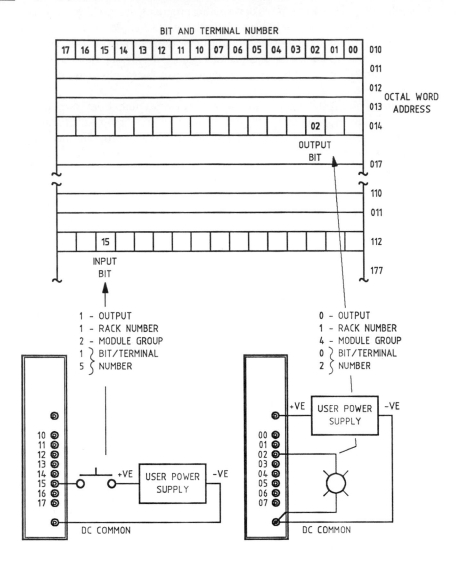

Figure 3.5   Relationship between input and output terminals and data table

DC inputs and outputs and their relation to their word and bit address are shown in Figure 3.5. The input or output assembly terminal number corresponds to the bit address in the data table. A 1 appearing in the data table indicates an 'on' condition; a 0 in the data table indicates the 'off' condition. The number of input or output modules and racks that are used in the system is dependent on user requirements; an Allen-Bradley PLC 2/15 could contain up to eight module groups, therefore providing connections for 128 field input/outputs in one fully configured rack, and as few as 32 input/outputs in a rack containing one input module group and one output module group. Allen-Bradley programmable controllers are available with a variety of input/output configurations as well as various types of input/output modules.

The memory organization of the Gould Modicon Micro 84 allows any mix of input/output modules with up to sixty-four input/output points in any single rack. It has a wide range of other types of input and output modules that can be fitted in the rack. It is available in two user memory sizes, either 1 k or 2 k depending on user requirements. The inputs and outputs are addressed by a four-digit number. The input address commences with the digit 1, and the outputs commence with the digit 0. Figure 3.6 shows the input/output organization of the Micro 84 with 2 k of memory.

Discrete	1	0008	0007	0006	0005	0004	0003	0002	0001
Output Module	2	0016	0015	0014	0013	0012	0011	0010	0009
	3	0024	0023	0022	0021	0020	0019	0018	0017
	4	0032	0031	0030	0029	0028	0027	0026	0025
Discrete	1	1008	1007	1006	1005	1004	1003	1002	1001
Input Module	2	1016	1015	1014	1013	1012	1011	1010	1009
	3	1024	1023	1022	1021	1020	1019	1018	1017
	4	1033	1032	1030	1029	1028	1027	1026	1025
Sequencers		2108	2107	2106	2105	2104	2103	2102	2101
Numeric Data									
Input Registers						3104	3103	3102	3101
Internal		4008	4007	4006	4005	4004	4003	4002	4001
Registers		4016*	4015*	4010*	4013	4012*	4011	4010*	4009
		4021*	4023	4022*	4021	4020*	4019	4018*	4017
		4032	4031	4030	4029	4028	4027	4026	4025

*Output Registers

Figure 3.6   Gould Modicon Micro 84 input/output organization

The processor in the Modicon Micro 84 is 4-bit which provides for 14 8-bit words. Four words are allocated to field input registers and four words to field output registers. The Gould Modicon Micro 84 is one of the smaller programmable controllers in the Gould Modicon range. The field connections for an AC input and output are shown in Figure 3.7.

The Hitachi P 250E has 2 k of random access memory for user program which can be expanded by 2 k by using an additional memory card (printed circuit board). The Hitachi P 250E basic unit has 32 input circuits and 24 output circuits. The input addresses are prefixed with the letter X, the output addresses with the letter Y, and both are shown together within internal storage bits and counter and timer allocation as follows:

Inputs X200 – X232
Outputs Y0 – Y24
Internal Storage Bits R 000 – 1199
Timers/Counters 000 – 255 prefixed TD, SS, WDT, CU, or CL.

The relationship between field connections and input/output terminals for the Hitachi P 250 E is shown in Figure 3.8.

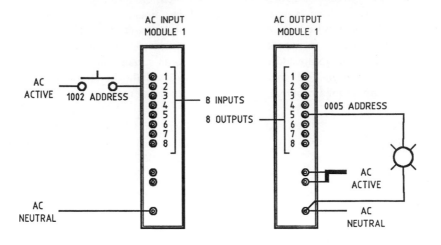

Figure 3.7   Relationship between input and output terminal and address

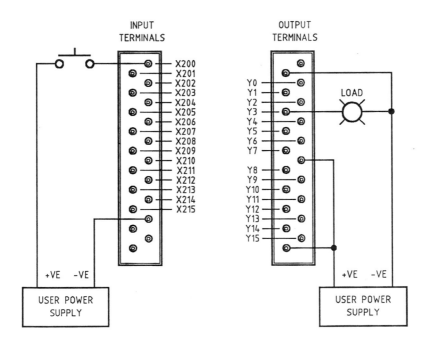

Figure 3.8   Relationship between input/output terminals and addresses

3.2 Flowcharts

A flowchart is a graphic method of representing a sequence of events. When it is applied to programmable controllers it is a method of obtaining a solution to a programming problem. This solution is sometimes termed an algorithm. Using flowcharts to design programs for programmable controllers leads to structured, well-defined programs, with the flowchart becoming the first step in producing a fully documented system. Flowcharts have the following advantages:

- They are independent of the type of system.
- They can be read and interpreted by people other than the programmer if standard symbols are used.
- They show the step-by-step order of instructions required to achieve the process.
- They show points in the program where decisions are required.

The flowchart is drawn as a block diagram using a set of standard symbols. It consists of a box which is appropriately labeled to indicate the step in the program that is being undertaken. The box is joined to others by lines with arrows to indicate the direction of the program flow. If a decision needs to be made, a box with a yes or no as alternative outputs will indicate the direction of program flow. The symbols used to produce programmable controller flowcharts are shown in Figure 3.9. A full list of flowchart symbols is shown in Appendix 2.

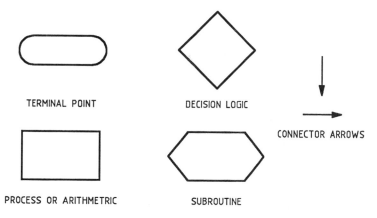

TERMINAL POINT   DECISION LOGIC

CONNECTOR ARROWS

PROCESS OR ARITHMETRIC   SUBROUTINE

Figure 3.9   Flowchart symbols

A flowchart is often used to subdivide the program into smaller, more easily programmed routines. An example of a simple flowchart is shown in Figure 3.10 where the flowchart has been written for a simple process with the following requirements:

1 There is to be a manual start sequence if no alarms are present and the process is not operating.
2 Two liquids are to be pumped into a container.
3 The liquids are to be mixed for one minute.
4 The liquids are to be pumped into a storage vessel.
5 The process is to be made ready for another start sequence.

41

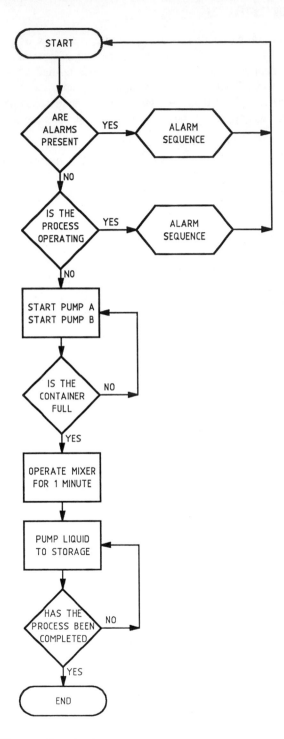

Figure 3.10   Process flowchart

Flowcharts have some disadvantages:

- They are time consuming to prepare and therefore costly to produce.
- They reflect one programmer's solution to a programming problem and, as there may be many solutions to any problem, another programmer may have difficulty in seeing the logic of the solution.
- They do not show some of the intervening steps in the program.

Flowcharts can be divided into two types: a straight line or sequential flowchart and a structured flowchart. The sequential flowchart is the type of chart most often used to assist preparing programmable controller programs. It can be used when the program does not have a large number of paths that can be followed; this is the case with most process control functions or machine control operations. The structured flowchart is most often used by computer programmers for complex multiple path computer operations.

Functional sequence charts

Manufacturers have introduced a sequential function chart programming facility for programmable controllers which allows the programmer to divide the process into steps that directly represent the function chart. The processor follows the function chart as it executes the program and will carry out a step only when the preceding step has been completed. An inactive step is not scanned. The steps of the process can be directly programmed as part of the sequence chart, or steps can be programmed as a set of conventional ladder diagram rungs. The use of the functional sequence chart simplifies the programming of the complex condition requirements and interlocking required in conventional ladder diagrams for an output action to occur. The function chart is entered directly from the programming terminal.

The advantages of sequential function chart programming are:

- It makes programming and logical program design easier.
- It reduces overall PLC program scan time by only scanning active program steps.
- It makes program fault location easier as only one part of the program is in operation at any given time.

An example of a sequential function chart is shown in Figure 3.11 where an automatic spray painting operation is taking place. If the process is not in operation, and no alarms are operational, an item can be painted any one of a number of colors. This sequence of operation is shown by the left branch of the sequence chart. After the painting operation has taken place, it is possible to return to the start of the program to spray another item, or end the program, depending on the function required. The right branch of the sequence chart shows the operation if a shutdown or system cleaning operation is required.

The symbols used in sequential function charts are as follows:

- ☐ is a step which may contain multiple rungs of ladder diagram logic;
- ✛ is a transition between two steps; this may be the location where decisions in the process are determined;
- ▬ is a multiple path where more than one operation starts or concludes;
- the step is indicated by a number, e.g. 002.

43

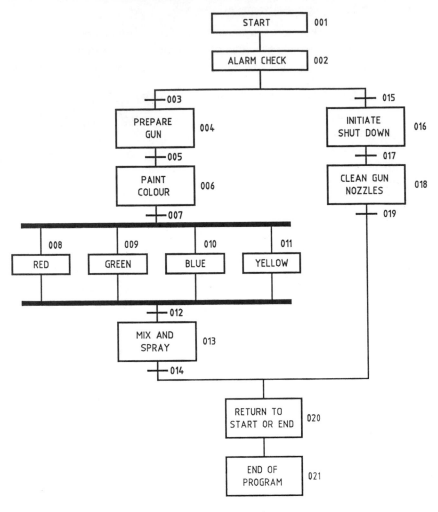

Figure 3.11   Sequential function chart programming

3.3 PLC ladder diagrams

Ladder diagrams have been extensively used for the documentation of electrical circuits. When programmable controllers were introduced, a simple programming technique was needed that required minimal programming training for the operator. The ladder diagram was selected to be the fundamental programming tool for programmable controllers as it was well-known in the electrical/electronics industry and could easily be adapted to a visual display unit used with the programmable controller.

The ladder diagram consists of two vertical lines. The left represents the positive or active supply, and the right, the negative, zero volt, or neutral of the supply. Between the supply rails, lines made up of relay contacts, contactors or solenoids are included. These lines represent the rungs of the ladder diagram. Figure 3.12 shows some of the electrical symbols used in ladder diagrams.

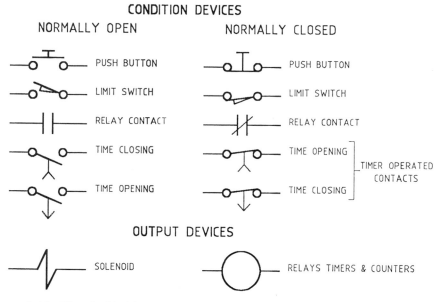

Figure 3.12   Electrical ladder diagram symbols

A ladder diagram, consisting of a limit switch, which must be closed, a pressbutton, which must be released, and a normally open relay contact, used to operate a solenoid, is shown in Figure 3.13 (a). The three contacts and the solenoid make up a single rung of a ladder diagram. The left side of the rung, where the limit switch, pressbutton, and relay contact are located, is termed the condition area of the rung. The three contacts must be in the correct condition to produce an output. The output or right side of the rung shows the device that will be made active when the input conditions are set correctly. Figure 3.13(b) shows the same diagram in PLC ladder diagram format.

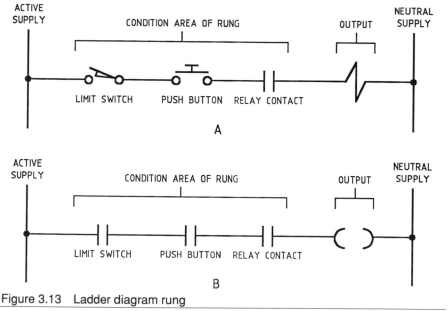

Figure 3.13   Ladder diagram rung

3.4 Ladder diagram design examples

Figure 3.14 is a simple process used as an example to demonstrate how a relay logic system can be developed to provide a control function. The diagram is used to develop an operational relay ladder diagram.

Example 1

A pump is used to fill two storage tanks. The pump is manually started by the operator. When the first tank is full, the circuit must be able to automatically fill the second tank by opening a valve which is actuated electrically. When the second tank is full, the pump must automatically shut down and a 'tank full' indicator lamp will light.

In Figure 3.15, rung 1 is the start/stop circuit used to initiate the operation. The stop button is normally closed and will open to operate. Operating the stop button will release the start relay. Releasing the start relay will stop the pumping operation. To start the pumping operation the conditions which must be met are:

1 that the stop button is not operated;
2 that the float switch 2 is not operated;
3 that the start button is operated.

Float switch 2 is the tank full switch. The system must be stopped by this normally closed switch opening when both tanks are full.

When the start relay is operated, a start relay contact closing bypasses the start button. This contact holds the start relay operated so that the operator does not need to hold the start button. The second and third start relay contacts prepare the second rung for operation. A fourth start relay contact supplies power to the pump in rung 5. The second rung controls the valve between the two tanks. The valve actuator relay will operate and open when:

1 the start relay is operated;
2 the float switch on tank 1 is not operated;
3 the float switch on tank 2 is not operated.

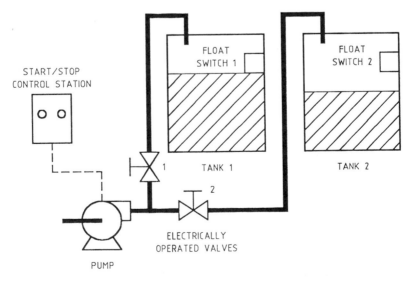

Figure 3.14   Example 1 system diagram

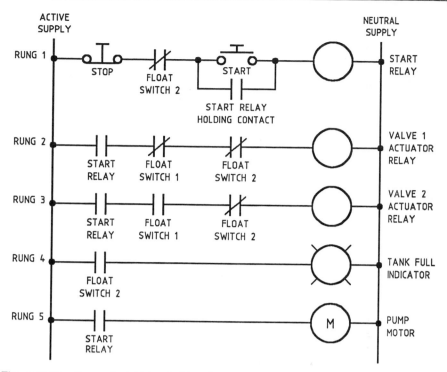

Figure 3.15 Example 1 Relay ladder diagram

When the float switch on tank 1 operates, indicating the tank is full, rung 2 is de-energised. This causes valve actuator 1 to close and energise rung 3, which opens valve actuator 2. Valve actuator 2 will operate when:

1 the start relay is operated;

2 the float switch 1 on tank 1 is operated;

3 the float switch 2 on tank 2 is not operated.

The fourth rung will light the 'tank full' lamp when float switch 2 on tank 2 operates. The system is stopped by releasing the start relay when the float switch 2 contact in rung 1 is opened by the tank being full.

This is only one solution to the operational sequence of this system; there will be others. It depends on system requirements and the individual circuit designer which solution is taken.

In the solution provided in Example 1, each rung cannot operate unless a set of specific conditions are met. The conditions for rung 2 operation, for example, require the start relay and float switch 1 to be operated, and float switch 2 to be operated. The condition in rung 2 interlocks the operation of the valve actuator to other conditions elsewhere in the system. Interlocks in ladder diagram programming are essential to safe operation of the system. They are included in the condition area of the rung.

When the condition area of a rung causes the output device to become active, the rung is described as being true and is considered to be passing power. When the rung is not active it is described as being not true (or false), and is not passing power. True and not true (or false) are terms often applied to logic and digital-type circuits; because

47

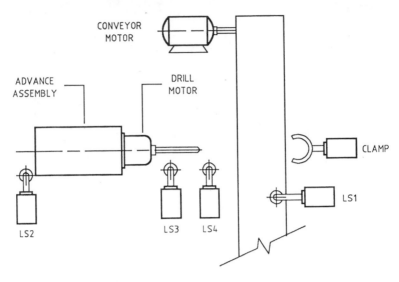

Figure 3.16   Conveyor drill system

a rung is a form of logic, the terms also apply. A contact or condition within a rung can also be described as true or not true, depending on its state of operation.

Example 2

This is a more complex system (see Figure 3.16) where a conveyor carries work which is to be drilled. When the operator operates the start button and places work on the conveyor, the system goes through the following sequence:

1 The conveyor starts, and carries the first item to be drilled until it strikes limit switch 1.
2 When limit switch 1 is operated the conveyor stops, and the item is clamped.
3 The drill is moved forward. At limit switch 3, the drill motor starts.
4 At limit switch 4 the drill has reached the required depth, the drill returns to limit switch 2, and is switched off.
5 The conveyor motor starts, and the same sequence is repeated.

Figure 3.17 shows the ladder diagram which is a possible solution for Example 2. The operation of the conveyor drill system is as follows:

Rung 1: This is the manual start stop system including the start relay holding circuit.
Rung 2: When the start relay is operated the conveyor motor will operate until an item to be drilled operates limit switch 1. Another condition that will stop the conveyor is if limit switch 2 is operated indicating the drill is not in the home position.
Rung 3: A contact of the motor start relay operates the conveyor motor.
Rung 4: When limit switch 1 is operated and the conveyor motor has stopped, the clamp solenoid will operate.
Rung 5: When an item is present (limit switch 1 operated), limit switch 4 is not operated, and the job complete relay is not operated, the advance drill solenoid will operate causing the drill assembly to move off limit switch 2.

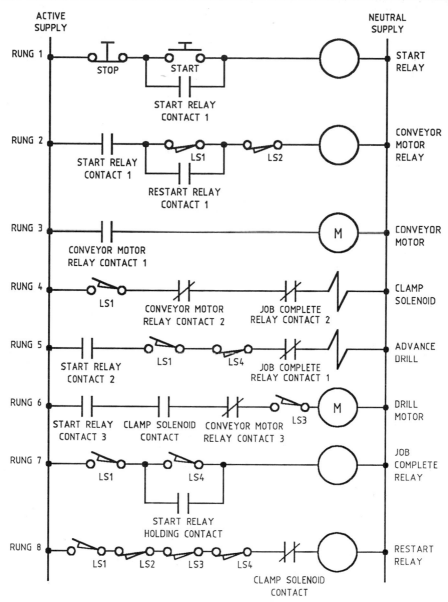

Figure 3.17   Example 2 ladder diagram

Rung 6: When an item is clamped, the conveyor is not operating, and the drill assembly is advanced sufficiently to cause limit switch 3 to operate, the drill motor will start.

Rung 7: When a job is clamped (limit switch 1), and the drill assembly has reached the full length of its travel (limit switch 4), the job complete relay will operate and hold itself operated via its own contact. The job complete relay releases the advance drill solenoid (rung 5) allowing the drill and the clamp solenoid to return to normal.

Rung 8: The restart relay had been held released due to the limit switch and clamp solenoid interlocking.

As the job complete relay is held in (rung 7) the drill assembly will return to normal causing the conveyor to operate (rung 2) via limit switch 2 which would be operated when the drill assembly has returned to the rest position.

As the conveyor motor is allowed to restart, the clamp solenoid (rung 4) is being released, allowing LS1 to become non operated. LS1 also releases the job complete relay in rung 7, allowing the system to be ready to receive another item on the conveyor to be drilled.

3.5 Programmable controller relay instruction set

There are a number of programmable controller instructions which are similar in function to relay logic and can be used to produce similar operations to the electrical ladder diagrams shown in the previous two examples. The PLC relay ladder diagram functions which will be considered are:

- examine instructions;
- output instructions;
- master/zone control relay instructions;
- branch instructions;
- latch instructions.

There are two examine instructions and an output instruction in the programmable controller relay instruction set:

1 $\dashv\vdash$ examine an input for an 'on' condition or a field device for a closed condition;
2 $\dashv\!\!\!/\!\!\!\vdash$ examine an input for an 'off' condition or a field device for an open condition;
3 $\dashv(\ )\vdash$ output.

The examine instructions examine the status of a bit located in the input data table. The status of the data table bit is a reflection of the 'on' or 'off' condition of a two-condition device connected to the input module of a programmable controller. When the device is 'on', the input is said to be true and a 1 will be the result in the data table. When the device is 'off', the input is said to be false and the result is a 0 in the data table. These conditions apply when the examine 'on' instruction is in operation. The examine 'off' instruction is true when the two-state input device is 'off' causing the data table bit to be set to 1. When the two-state field device is 'on', the examine 'off' instruction is said to be false, causing the data table bit to be set to 0.

Output instructions are programmed at the right-hand end of a rung. An output instruction is usually preceded in the rung by some condition logic. When it is not, the output is unconditional and is permanently 'on', or energized. When an output becomes energized, the condition logic is such that a logical path for electrical continuity exists across the rung. For example, a particular output instruction, when active, sets a bit in the table to a logical 1, and, when not active, to a logical 0. If the output bit is connected via an output module to a physical output and the rung is true, the output is activated. As there is a limitation on the number of physical inputs and outputs that can be connected to a programmable controller, some unused data table bits can be used as storage points. Bits used for storage can act as internal relays. These are of assistance when programming as any bit in the data table can be examined any number of times for a true or false state.

Another type of output instruction is the latch, the symbol for which is $-(L)-$. There are many applications that require the output, once operated, to remain in the operated state until an unlatch signal is indicated by the user program. When a rung containing a latch is true, the latched output will remain 'on' after the rung condition becomes false. A latched output will return to its pre-power fail state if power is removed from the programmable controller and then returned; that is, a latched output will not change its state unless deliberately unlatched. The symbol for unlatch is $-(U)-$.

The unlatch command, when true, causes a latched output to become unlatched providing the latch rung has become false. The latch and unlatch command are termed set and reset respectively in some programmable controllers. The set command will cause one output to latch, the reset command addressed to the same output will cause the output to unlatch providing the condition of the set rung is false.

Master control relays are used in relay logic to allow control of a number of rungs or an entire circuit by a single relay operation. In relay circuits the master control relay disconnects power to the parts of the circuit that are to be inhibited when the master control relay is not operated. A similar function has been included in many programmable controllers. When a rung containing a master control relay function is true the program will function, each rung within the master control relay area operating normally. When the master control relay rung is false the outputs associated with the master control relay will release. The rungs within the master control relay area are defined by a master control relay or set command at the start of the controlled rungs, and another master control relay or reset command at the end of the controlled area. The symbol is $-(MCR)-$.

Allen-Bradley offer a zone control function. The symbol for this is $-(ZCL)-$. The outputs of the rungs within the controlled zone will remain in the last state when the zone control rung at the start of the controlled zone is false. When the zone control function is true the rungs within the controlled area operate normally. The end of the controlled zone is defined by another zone control function.

Branching

Parallel connections for branches within a user program require a branching or interconnection instruction. The Allen-Bradley programmable controller uses a symbol for branch start and another for branch end (shown in Figure 3.18) at the start of a parallel connection. The Modicon Micro E84 and the Hitachi P 250E use the interconnection symbols to achieve parallel connections. The interconnection symbols are shown in Figure 3.19(a), where a vertical interconnection is made, and 3.19(b) where a horizontal interconnection is made.

A   B

BRANCH START/OPEN   BRANCH END/CLOSE

Figure 3.18(a) Allen-Bradley branch start   (b) branch end symbols

A
VERTICAL LINK

B
HORIZONTAL LINK

Figure 3.19   Rung interconnection symbols

The number of series and parallel relay functions that can be programmed into an array is limited and varies with programmable controllers. Allen-Bradley allow a maximum of eleven series functions and up to seven parallel branches. Any rung network can operate one output which will be located on the first rung of the program. Modicon Micro E84 allows ten series elements and up to seven parallel branches. Any combination of elements can operate up to seven parallel outputs. The Hitachi P 250E allows nine series elements and up to seven parallel branches to operate one output.

Programming more than the allowable series elements or parallel branches will result in an error message being displayed.

If a program requires more series elements than can be programmed into a single rung, a storage bit is used to produce the desired logic. In the example shown in Figure 3.20, where fourteen series elements are required to produce an output, a storage bit located at address 02000 is used as a condition to produce an output in the second rung.

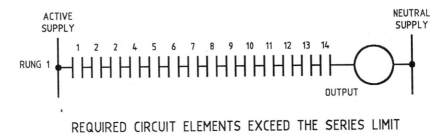

REQUIRED CIRCUIT ELEMENTS EXCEED THE SERIES LIMIT

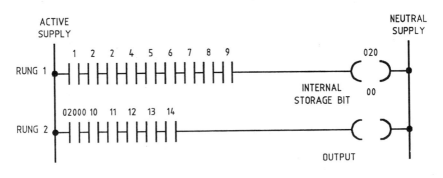

PLC CIRCUIT

Figure 3.20   Storage bit used to obtain required logic

Nesting

Some programmable controllers allow nesting of inputs. A nest of inputs is a branch within a branch and is allowed as part of the user program. Programming nested contacts makes the programmable controller program appear closer in layout to an electrical ladder diagram. This makes it easier for maintenance personnel because they are familiar with electrical ladder diagrams. The Hitachi P 250E and the Modicon Micro E84 both allow nesting of program elements; the Allen-Bradley PLC 2 series does not. Allen-Bradley have had the nesting function included in the PLC 5 family. Figure 3.21(a) shows an electrical hardwired diagram where nested elements are included. An equivalent program suitable for a programmable controller is shown in Figure 3.21(b).

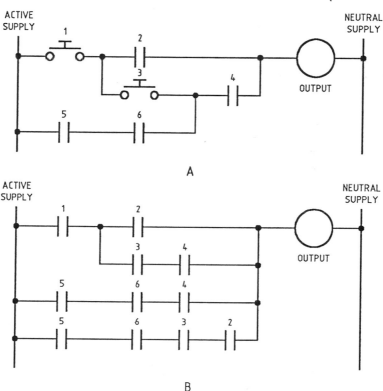

Figure 3.21   Nesting relay instructions   (a) Electrical ladder diagram   (b) PLC ladder diagram

3.6 Programmable controller program examples

To place a user program into a programmable controller memory, the programmable controller has to be made ready to accept the program. Programmable controllers have various modes of operation, and the correct mode must be selected to enter a new program. When a new program is entered, the program write, program or enter mode must be selected. The name used to describe the modes varies between manufacturers. The operating modes of the Allen-Bradley PLC2, Gould Modicon Micro 84, and Hitachi P 250E are shown in Table 3.1.

Table 3.1   Operating mode names

Allen-Bradley	Modicon	Hitachi
Program	Examine	Write
Remote program	Data entry	Read
Run/program	Supervisory	Force
Run	Error	Monitor
		Insert
		Delete
		Run

A separate switch, often key-locked, is provided on most programmable controllers to switch the system to the run mode. The run mode, when selected, allows the user program to operate. The run switch is included so that when the programmable controller is connected to a machine or process a deliberate action is required to make the program execute. The provision of such a switch provides some degree of safety for personnel and machinery.

Prior to entering a new program it is advisable to clear the programmable controller memory. To clear the programmable controller memory the following keystrokes are required:

- For an Allen-Bradley with a T3 industrial terminal, select the type of programmable controller from the mode selection menu displayed on power up. Switch the programmable controller to the program mode and press

CLEAR MEMORY	9	9

- For a Modicon Micro 84 with a P370 programming terminal, after first stopping the CPU (supv 1) select the supervisory mode 3 and press the ENTER button twice.
- For a Hitachi P 250E with a hand programming terminal, select the auxiliary mode and press

AUX	↓	↓	ENT	STA	RD/WRIT

In the following program examples, it is assumed that input and output assemblies are available in the appropriate locations to obtain a result when the program is operated. The examples shown can be programmed into any PLC providing the program requirements are met according to the manufacturer's handbook. To test the program's operation, the programmable controller must be switched to the run mode.

Warning: All programming examples in this book are designed to operate and should only be programmed into a PLC with an input/output simulator. If examples are programmed into a PLC connected to a machine or process, injury to personnel or damage to equipment may result. A simulator design is shown in Appendix 6.

Examine 'on' and 'off' instructions

The operation of the examine 'on' example in Figure 3.22(a) is as follows. Switch address 11000 is operated causing a field contact closure. Power flow through the rung will occur and the output at address 01100 will operate. The operation of the examine

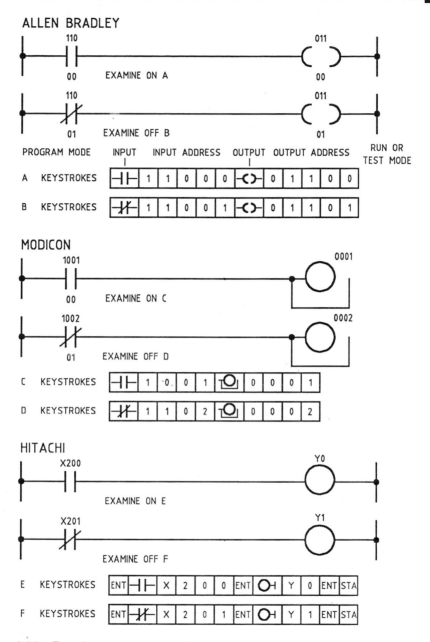

Figure 3.22 Examine program examples

'on' circuits in Figure 3.22(c) and (e) are similar in operation.

The operation of the examine 'off' example in Figure 3.22(b), (d) and (f) is that when the switch is operated, the rung will cease to have power flow and the output will become inactive.

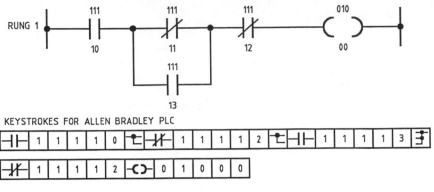

Figure 3.23   Programming branches

Branching

As there is an obvious similarity between the programming of the different types of programmable controller (see Figure 3.22), the following relay instruction examples will show one program to cover all of them; actual address numbers will be omitted. Figure 3.23 is an example of relay instructions where branching is included. The keystrokes necessary to program branching instructions are listed; each instruction is located in a separate box containing the programming information. This method of designing and documenting a program is satisfactory for small uncomplicated programs only.

Programming stop/start circuits

The electrical stop/start circuit shown in Figure 3.24(a) consists of a normally closed stop button in series with a normally open start button, with a holding contact of the start relay in parallel with the start button to hold the start relay operated when the start button is released. When this circuit is programmed into a programmable controller, the following conditions must apply to cause the start relay to operate:

1 the start switch must be examined for a closed condition;
2 the stop switch must be examined for a closed condition.

Because both the start and stop button must be in a closed condition to cause the start relay to operate, the start and stop buttons are programmed in exactly the same way, as shown in Figure 3.24 (b).

Warning: As a safety measure, emergency stop pressbuttons should never be derived as part of a programmable controller program but should be hardwired to remove power to the machine or device when they are operated. The emergency stop can be used within the PLC program but the user program should never be relied on to stop a machine or process in an emergency situation.

Programming a latch

The ladder diagram shown in Figure 3.25 is an example of a latch and unlatch for the Hitachi P 250E. The latch command is achieved by using the set and the unlatch reset.

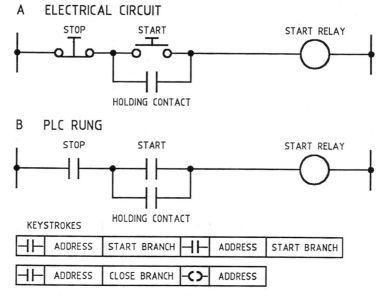

Figure 3.24   Programming start/stop circuits

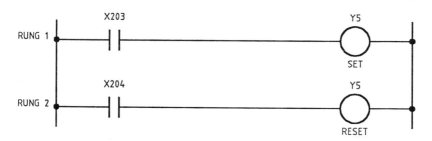

Figure 3.25   Programming latch and unlatch

The circuit operates and latches output Y5 when rung 1 is true via X203. The output Y5 will remain latched independently of X203 being operated. To unlatch output Y5 rung 1 must be false and rung 2, the reset rung, must be true via X204.

3.7 Designing programmable controller programs

The simple one or two line programmable controller programs discussed earlier require little design consideration and can be programmed directly into the programmable controller without difficulty. As programs become more complex, however, a method of program design becomes necessary. A flowchart is a method of determining the system objectives that must be met by the program and is a starting point in program design. A flowchart, however, does not show the individual program elements and their relationships within the program.

There are four steps in developing a new or modifying an existing programmable controller program. These steps are:

1 If the programming required is related to a modification, analyze the existing system and be sure its operation is understood.
2 Write a list of objectives that will describe what is being achieved by the required system.
3 List the hardware and software to meet the requirements in 2 and make a block diagram or flowchart sketch of the system operation.
4 Produce a decision table to meet the requirements of the block diagram or flowchart.

A decision table, sometimes termed a state table, consists of a cross reference between the input element condition, and the output element condition. A symbol is used to indicate the state of an element within the rung. The symbols used in this programmable controller decision chart are:

O = operated element
O̅ = not operated element

The bar across the top of the O indicates the logical NOT condition.

R = ready for operation
R̅ = not ready for operation

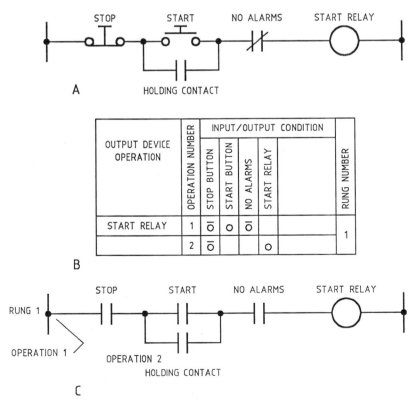

Figure 3.26   Programmable controller program development

Figure 3.26(b) is a decision chart for the start sequence shown in the electrical ladder diagram in Figure 3.26(a). From the decision chart, the programmable controller ladder diagram is derived as shown in Figure 3.26(c). The symbol O means that the element in the rung associated with the symbol must be operated in order to obtain power flow through the rung to cause the output to become active. The $\overline{O}$ is the symbol used to indicate that the condition element is not operated to pass power to cause an output to become active.

The R and $\overline{R}$ used in the table are for programming counters and timers. The R indicates that a timer or counter is ready to pass power or, in a preset state, to cause an output to become active. The $\overline{R}$ indicates that the counter or timer has not reached a ready or preset condition to pass power to allow an output to become active.

An unused square on the table indicates that the input device associated with the square plays no part in the logical operation of the output element.

In the rung of programming shown in Figure 3.26(b), three conditions are required to cause the start relay to operate.

- the stop button must be not operated ($\overline{O}$);
- the start button must be operated (O);
- the no-alarms relay must be operated ($\overline{O}$).

The parallel holding circuit around the start pressbutton must be complete to obtain permanent operation of the start relay. The hold in operation is the second operation of rung 1:

- the stop button must be not operated ($\overline{O}$);
- the start relay must be operated (O).

As more complex programs are programmed into a programmable controller it is recommended that a program development method, such as a decision chart, be used to produce a logical and well-documented program.

3.8 Test questions

These questions relate to programmable controller programs and programming. It is suggested that, in addition to producing the solutions to the problems on paper, the solutions be programmed into a PLC and tested. Hands-on experience is invaluable for a full understanding of programmable controller operation.

The solutions to the odd-numbered programming exercises are in the answer section of this book. The solutions are similar for all types of programmable controllers, with the exception of the input and output assembly addresses. Even-numbered solutions are included in the teacher's manual.

1 Draw a flowchart for the process shown in Figure 3.27, where steam is used to heat a liquid to 70°C. A level switch is located in the vessel to make sure the water level is maintained. Liquid cannot leave the vessel until the liquid temperature is 70°C. The steam is to shut off automatically if the liquid become excessively hot.

2 What is the relationship between the following addresses of the inputs and outputs and their physical rack module location?

Allen-Bradley	(a) 01111	(b) 11112
Modicon	(a) 0010	(b) 1001

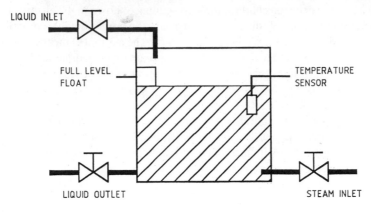

Figure 3.27   Question 1

3 Draw a programmable controller ladder diagram for the circuit in Figure 3.28.

4 Redraw the contact cluster in Figure 3.29 so that it could be programmed into a programmable controller which has a 9 × 8 input/output network limitation.

5 Redraw the electrical contact cluster shown in Figure 3.30 in programmable controller format so that it could be programmed into a programmable controller that does not allow contact nesting.

6 List the keystrokes necessary to program the ladder diagram in Figure 3.31, including address keystrokes.

7 Draw a decision chart and ladder diagram for the electrical circuit shown in Figure 3.32.

8 Describe in words the operation of the Allen-Bradley PLC circuit in Figure 3.33.

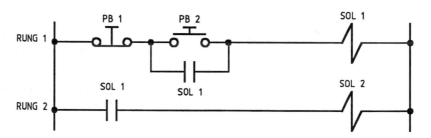

Figure 3.28   Question 3

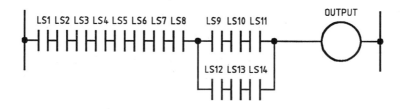

Figure 3.29   Question 4

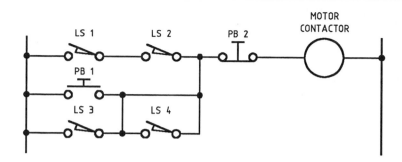

Figure 3.30   Question 5

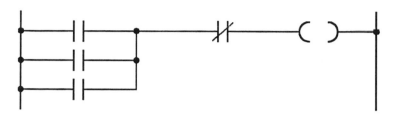

Figure 3.31   Question 6

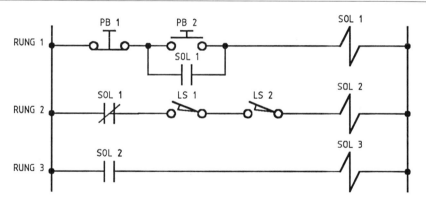

Figure 3.32   Question 7

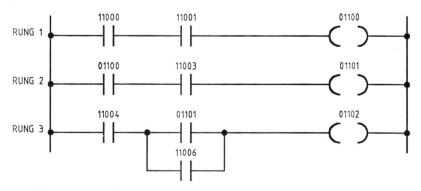

Figure 3.33   Question 8

CHAPTER FOUR

System operation and logic programming techniques

4.1 PLC systems

The programmable controller is not a separate device operating in isolation; even in the simplest application it is the central section of an overall control system.

The programmable controller reduces costs and increases efficiency by:

- replacing discrete timers and counters;
- performing arithmetic functions;
- controlling process loops;
- replacing analog electronic circuits;
- providing inexpensive documentation;
- interfacing to computers;
- performing position and speed control functions;
- operating in an industrial environment;
- controlling factory-wide automatic control;
- operating multiple programs or controlling more than one machine;
- detecting process errors and providing operational alarms.

4.2 Modes of operation

All programmable controllers have at least several modes of operation. These modes can be divided into three major categories with several sub modes in each category. The three major modes of operation are program run, program edit, and program write. In the program run mode the controlled process or machine is allowed to operate. The inputs and outputs are active and the user program, in conjunction with the inputs, is being scanned by the processor to cause various outputs to operate.

In the program edit mode, the user can make alterations to the program. Two separate types of editing are available on most programmable controllers: on line, and off line. Off line editing allows alterations to be made to the user program while the programmable controller is not scanning the inputs, outputs, or the user program. On line editing allows the user program to be edited while the controlled outputs are in operation. Extreme care must be taken when on line editing functions are being carried out as program alterations may cause incorrect or hazardous operation of the controlled device or process, possibly causing injury to personnel or damage to machinery.

In the program write mode, the user program can be entered. The write mode stops the operation of the program and inhibits the operation of the outputs. A test mode is included in some programmable controllers where the user program can be tested, but, like the program write mode, the outputs are inhibited.

An explanation of each mode of operation for the Allen-Bradley PLCs, Gould Modicon Micro 84, and the Hitachi P 250E is included to provide the PLC user with an overview of the terminology and operation of most operational modes found in programmable controllers.

Allen-Bradley PLC 2 modes

The Allen-Bradley PLC 2/15 has a key switch located on the processor module to switch the processor to one of three modes of operation. These modes are run, test, and program.

In the run position, the user program will be executed, and field input information, together with the user program, will cause outputs to be operated. When in the program mode the user program cannot be changed, but data table alterations can be carried out using an on line data change function or bit manipulation technique (discussed in Chapter 8). The user program can be tested in the test mode. The inputs to the programmable controller are active and will cause the user program to respond accordingly; the outputs are not activated, but are disabled when this mode is selected.

The program position allows user program instructions to be entered. The user program can be manually entered by the operator or entered via a form of mass storage such as magnetic tape, computer, discrete memory chips, etc. When in the program mode the PLC will not allow outputs to become active and therefore stop the operation of the process or machine.

Other Allen-Bradley PLCs have some slightly different or additional modes of operation. The PLC 2/17 has an additional mode of operation termed remote program. The remote program mode of operation allows program changes to be made from a location some distance from the processor via the serial communications port.

Gould Modicon Micro 84 modes

The Micro 84 programmable controller is interfaced with the P 370 programmer as shown in Figure 4.1. The Modicon P 370 programmer has four operating modes: examine, data entry, supervisory, and error.

The programmable controller enters the examine mode as soon as the power up diagnostics have been completed successfully. The examine mode is the monitoring mode of the system and allows the user program to be executed. The operational status of any system element can be determined while in this mode. The HOLD button, when used in this mode, will cause the contents of the data display to be frozen to allow examination of reference values. The data entry mode allows the user program to be entered into the controller memory. When the ENTER button is pressed the data displayed on the programmer is entered into the processor memory. The supervisory mode allows the user to initiate the supervisory commands shown in Table 4.1. The supervisory key is abbreviated on the controller as SUPV. The SUPV key is associated with a code key to select the desired command.

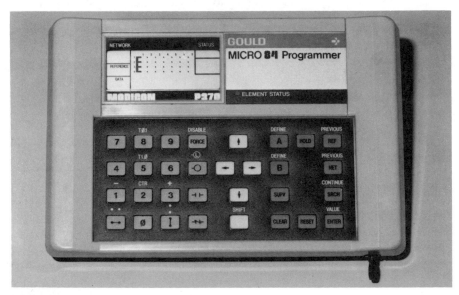

Figure 4.1   Gould Modicon P 370 programmer

Hitachi P 250E modes

The Hitachi P 250E has two fundamental modes of operation: run, and stop. A contact closure initiates the run mode which allows the memory to be scanned, and executes the user program providing no program errors are apparent. When the run contact is not operated (open circuit) the stop mode is initiated and program execution ceases.

The run/stop modes of operation are further subdivided into four sub modes: programming, monitor, auxiliary, and interface. The programming mode is selected by the operation of the read (RD) or write (WRIT) button. When the WRIT button is selected, the user program can be inserted, deleted or changed. The RD button allows circuit rungs and addresses to be displayed.

The monitor mode is selected by operating the MON button. The monitor mode allows the on/off status of circuit elements to be displayed. It also allows the operator to force an output on or off for testing purposes, and to change the time base values of timers and counters while the program is being executed.

Table 4.1   Modicon Micro 84 supervisory codes

Command	Code	Operation
Exit	0	Programmer is returned to the examine mode
Stop	1	Stops the program scan and retains outputs on/off status
Start of	2	Starts the program scan with the on/off status outputs prior to stopping the controller
Clear Memory	3	Allows the user section of the processor's memory to be erased providing the controller is stopped by using SUPV 1
Real Time	4	Display the logical condition of each rung as it is updated during the program scan
Dump Memory	5	Initiate the user program to be placed into an external program pack

In some instances the ENTER button must be pressed twice to carry out a command. THe first operation allows confirmation of the action, the second carries the action out.

The auxiliary function allows checks for:

- syntax errors;
- timer counter multiple use;
- definition number multiple;
- master control rationality;
- check sum value.

In addition, the auxiliary function allows:

- display of remaining free memory;
- clearance of user program;
- magnetic tape cassette interface for loading, recording and verification of programs;
- global change of input/output numbers;
- the defining and display of program areas to be retained at power failures.

The interface function is available to allow the connection of a printer for hard copy of ladder diagrams and to load the user program into read only memory (ROM).

Basic PLC system

Figure 4.2 shows a PLC configuration where the programmable controller may be dedicated to the controlled machine or process, or the machine or process may be reconfigured to carry out more than one function which would require a change of programmable controller program. External devices connected to the programmable controller, such as the printer, are termed peripherals and will be discussed in Section 4.4 of this chapter.

65

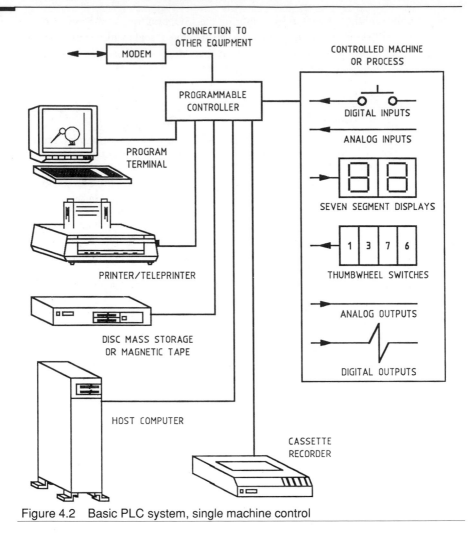

Figure 4.2   Basic PLC system, single machine control

Multitasking

Programmable controllers can be used in systems where a single PLC may control more than one machine or processor simultaneously. Multiple machine control is termed multitasking. PLCs used for multitasking are usually of the larger variety with more than 128 input/output connections. The multitasking arrangement reduces flexibility in machine and control applications because the machines and hardware features, once selected, are designed to suit a single application or installation. A further disadvantage of the multitasking system is that when the PLC requires down time (inoperative time) for purposes such as program changes or system maintenance, the total system ceases to function because the PLC is the central controlling element of the system. PLC down time could prove extremely costly on an installation such as an automotive production line.

The advantage of multitasking PLCs is that once they are installed and operating, maintenance and fault finding are easier and program modification is simple. Figure 4.3 shows a block diagram of a multitask programmable controller system.

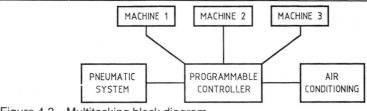

Figure 4.3   Multitasking block diagram

Distributed control systems

Figure 4.4 shows a block diagram of a distributed programmable controller system which consists of several programmable controllers, usually relatively small in input/output size (less than 128 I/O), connected together via a data communication network. To connect and access the data network, a modulator/demodulator unit, termed a modem, is used to encode and decode the digital signals which allow communication (see Chapter 10).

The complete system can be supervised by a host computer which would handle program up and down load, alarm reporting, data storage, and operator interaction facilities. The host computer in many instances would be a personal computer. Remote computer operators overseeing the system can be located some distance from the industrial environment in dust-free and temperature-controlled conditions. Each programmable controller in the system would control its associated machine or process which would mean that in many instances one programmable controller failure would not stop the complete system.

If multiple distributed control systems are located in an application, a process control computer, mini or mainframe, may be used to control the individual system computers.

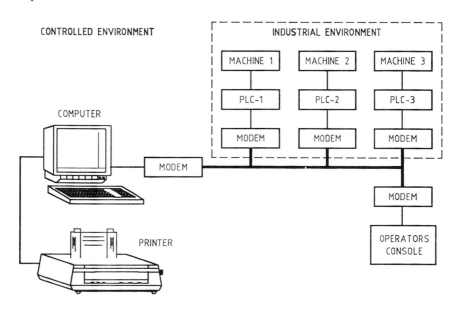

Figure 4.4   Distributed control systems

4.3 Program scanning

When the programmable controller is switched to the run mode the program is scanned in a cyclic manner. The program cycle is made up of an input scan, to detect input condition changes, a user program scan to solve the user program logic, and an output scan to set output conditions to the required state. The user program is scanned and usually executed in the same order as the program commands were entered by the user. There are, however, several other program scanning techniques used by various manufacturers which relate to the type of microprocessor used in the processor section of the programmable controller. The common program scanning techniques used are:

- from the first command, top left in the program, horizontally, rung by rung;
- from the top left command entered in the ladder diagram, vertically, column by column.

Parts of the user program can be removed from the scan by using a programming function termed jumps (see Section 4.7).

The time taken to scan the user program is dependent on the clock frequency of the microprocessor system. The higher the clock frequency, the higher the scan rate. A typical scan rate per instruction is 25 microseconds. Given that 1000 instructions is the length of a typical program, it takes approximately 25 milliseconds to scan the user program. This time does, of course, vary greatly between PLC manufacturers. As the scan time of the user program is dependent on the length of the user program, that is, the number of program instructions, the user program scan time can vary from as short as a few microseconds to in excess of 100 milliseconds.

Input/output scan

The full scan of a programmable controller must include inputs and outputs. The time taken to scan 128 of them is approximately 1 millisecond. The scan cycle normally begins with inputs, followed by the user program scan where the program algorithms are solved, followed by outputs. As there is a finite time difference between the operation of a field input and the resultant operation of the required output, compounded real time run errors can result. In most PLC applications the speed with which the program is executed is sufficient to cause no problems, but in high speed applications, where the program scan time is significant, steps must be taken to ensure that the speed of the application does not exceed the speed of the programmable controller scan.

Immediate instructions

An immediate input and output instruction is provided in the instruction set of some programmable controllers. This instruction is programmed in the user program and, when true, interrupts the program scan and updates a word in the input or output image table. This instruction can be used where it is important to produce an output, or act on an input, as soon as possible after its real world occurrence. Real world functions are considered to be discrete field inputs and outputs.

4.4 Connection of peripherals

A peripheral is any device that can be connected to the programmable controller that is not an integral part of it. The peripheral device is therefore a support device (see Figure 4.2). The variety of peripheral devices is large and includes:

- printer to produce ladder diagram and other hard copy documentation for the system;
- teleprinter to be used as an alarm and system data logger;
- magnetic cassette recorder for mass storage of program material;
- disc drives for mass storage of system information, including user program;
- modem for interconnection of programmable controllers via a data communications link;
- programming device to allow the user to enter the required program into the processor memory.

The peripheral devices are connected to the programmable controller via a communications port. The information may be communicated to the peripheral by means of serial signal or a parallel signal. These signals are a form of data communication and will be discussed in Chapter 10.

In some cases, the programmable controller manufacturer uses nonstandard data communication signals which limit the user to peripheral devices supplied by the manufacturer.

4.5 Logic function relationships to ladder diagrams

The use of logic functions in programmable controller program development provides PLC users with a simple and effective programming tool. The logic functions are based on the binary counting system, which has only two digits, 0 and 1. The 0 is considered to be the 'off' condition and the 1 the 'on' condition. As the majority of inputs to the programmable controller are on/off devices, such as a limit switch, and output devices are relays, solenoids or contactors, the application of logical conditions to circuit operation can simplify programming. The logical conditions discussed here are:

- AND;
- OR;
- NOT;
- NAND;
- NOR;
- exclusive or (XOR).

The AND function can be represented by two switches and a lamp as shown in Figure 4.5(a). When switch A and B are operated, the output C, the lamp, becomes active and turns on. If the active state is considered to be a logical 1 and the inactive state a logical 0, a truth table can be developed for the AND function as shown in Figure

69

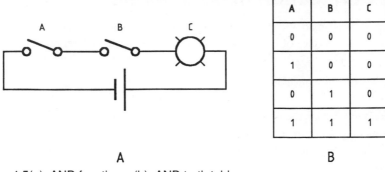

A	B	C
0	0	0
1	0	0
0	1	0
1	1	1

A   B

Figure 4.5(a) AND function   (b) AND truth table

4.5(b). When 1 is used to depict the active state and 0 the inactive state, positive logic is being used. In some applications the reverse applies where the active state is a logical 0 and the inactive state a logical 1; this is termed negative logic.

In Figure 4.5 the output C can only become active when switch A and B are operated. The AND ladder diagram function is shown in Figure 4.6(a) where contact A and B in the condition area of the rung must be operated to produce an output at C. The AND operation as a logic function has a symbol as shown in Figure 4.6(b).

The OR function can be represented by lamps and switches as shown in Figure 4.7(a). When switch A or switch B, or both switch A and B, are operated, output C is active. The truth table for the OR function is shown in Figure 4.7(b).

An OR ladder diagram is shown in Figure 4.8(a) where either of the parallel contacts A or B may be operated in the condition area of the rung to cause the output C to operate. The symbol for the OR function is shown in Figure 4.8(b). The AND and OR functions shown have two inputs. It is possible for these functions to have multiple inputs and they can be programmed accordingly.

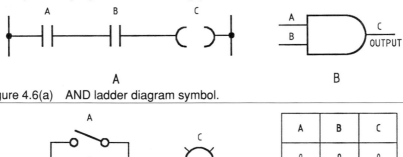

A   B

Figure 4.6(a)   AND ladder diagram symbol.

A	B	C
0	0	0
1	0	1
0	1	1
1	1	1

A   B

Figure 4.7(a) OR function   (b) OR truth table

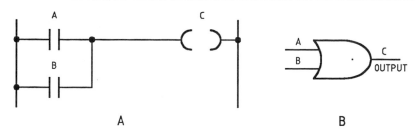

Figure 4.8(a) OR ladder diagram   (b) OR symbol

The NOT function is also known as an inverter in the logic family. The NOT function is often indicated by using a bar across the top of the letter indicating an inverted output. An inverter produces an output opposite to the input. The NOT function is shown in Figure 4.9(a) where a normally closed contact is in series with the output. When the contact is not operated the output is active and when the contact is active, the output is not operated. The ladder diagram and truth table of the NOT function and the inverter symbol are shown in Figure 4.9(b), (c), and (d). The small circle at the output of the inverter is termed a state indicator and indicates that an inversion of the logical function has taken place.

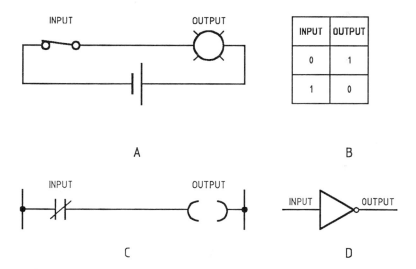

Figure 4.9(a) NOT circuit function   (b) Truth table   (c) Ladder diagram of a NOT function   (d) inverter symbol

The NOT AND, termed NAND, function is the logical AND operation followed by an inverter. The NAND function is often used in integrated circuit logic arrays and can be used in programmable controllers to solve complex logic. Figure 4.10(a) shows the ladder diagram circuit required to produce the NAND function. The only time that the output is not operated is when the storage bit is true. The storage bit is true when A and B inputs are active. The truth table and symbol for the NAND function are shown in Figure 4.10(b) and (c) respectively.

71

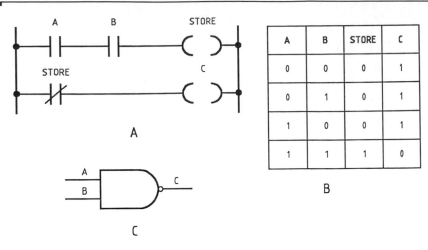

A	B	STORE	C
0	0	0	1
0	1	0	1
1	0	0	1
1	1	1	0

Figure 4.10(a) NAND ladder diagram   (b) NAND truth table   (c) NAND symbol

The NOT OR, termed NOR, function is the logical OR function followed by an inverter, and like the NAND function is often used to solve complex logic. Figure 4.11(a) shows the ladder diagram required to produce the NOR function. When A and B are not operated, output C will be true. When A or B, or both A and B, are operated, output C will not be operated (see Figure 4.11(b), the truth table for the NOR function). Figure 4.11(c) shows the symbol for the NOR function. A comparison of the NAND and the NOR truth tables will reveal that the logic produced is the opposite of one another.

If the input logic is inverted for the OR function, a NAND function is produced, thus producing the truth table as shown in Figure 4.12. The input logic is inverted by changing the ladder diagram inputs A and B from active closed contacts to active open contacts, that is, the normally open contacts are changed to normally closed. Using this technique to produce the NAND function eliminates the storage bit. Inverting the input

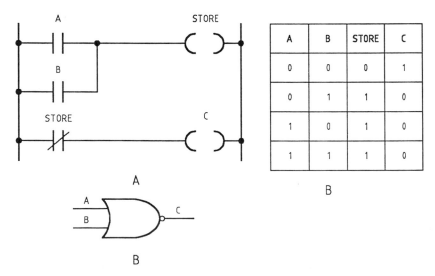

A	B	STORE	C
0	0	0	1
0	1	1	0
1	0	1	0
1	1	1	0

Figure 4.11(a) NOR ladder diagram   (b) NOR truth table   (c) NOR symbol

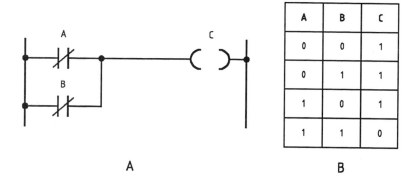

A	B	C
0	0	1
0	1	1
1	0	1
1	1	0

A   B

Figure 4.12   NAND function using inverted input logic

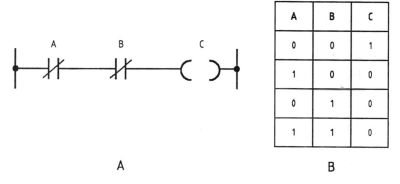

A	B	C
0	0	1
1	0	0
0	1	0
1	1	0

A   B

Figure 4.13   NOR function using inverted input logic

logic for the AND function produces a NOR function (Figure 4.13). Inverting the input function means that the logic is effectively a NOT function at the input of the logic array rather than being at the output as the NAND and NOR functions just described were. The state indicators would therefore be at the input to the symbol and the symbol would represent the true logical function as shown in Figure 4.14.

The exclusive OR (XOR) is true only if the inputs A and B are opposite logic conditions. If inputs A and B are both true or both false the exclusive OR function is false. Figure 4.15(a) shows the ladder diagram required to produce the exclusive OR function. The operation of the ladder diagram is such that at any time A or B are operated

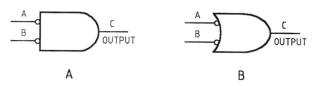

A   B

Figure 4.14(a)   NAND function   (b)   NOR function

73

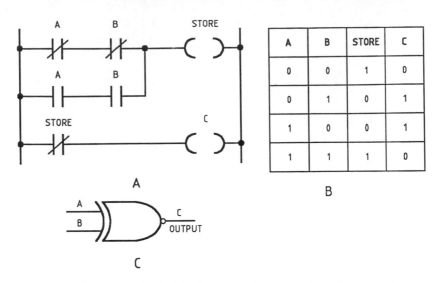

A	B	STORE	C
0	0	1	0
0	1	0	1
1	0	0	1
1	1	1	0

Figure 4.15(a) Exclusive OR ladder diagram   (b) truth table   (c) symbol

the storage bit is released and the output C is true. When A and B are both at logic 0 or 1, however, the storage bit is true and output C is false. The exclusive OR function and symbol are indicated in Figure 4.15(b) and (c) respectively.

4.6 Programming using Boolean algebra

Boolean algebra is a mathematical logic system based on two states, true and false. The Boolean system uses the AND, OR, and NOT functions to produce logic expressions that can be applied to the logical solving of field contact configurations in programmable controller systems.

The AND function is represented by multiplication or a dot symbol between the elements of the function: A and B is written A·B or A × B. The OR function is represented by the addition symbol, thus A or B is written A + B. The NOT function is represented by complementation or by the bar across the symbol. NOT A is written A′ or $\overline{A}$.

Using Boolean representation for logical functions and knowing that the states represented by the functions are 1 and 0, mathematical processes such as multiplication, addition and complementation can be applied to simplify or solve equations. In some instances, though, the logic hardware will not be the same as its algebraic mathematical equivalent even though the correct logic function is achieved. There are many applications where Boolean algebra could be applied to solving PLC programming problems, and in fact some programmable controllers can be directly programmed using Boolean expressions.

The simple start/stop circuit shown in Figure 4.16 can be expressed as a Boolean statement. The logic for this circuit is that output C will occur if A is closed and B or C1 are closed; therefore A · (B + C1) = C. If the start/stop circuit can be written as a Boolean expression it can also be drawn using logic gate symbols as shown in Figure

4.17(a) and written as a truth table as shown in Figure 4.17(b). In the truth table a closed contact is represented by a logical 1 and an open contact by a logical 0.

A Boolean expression can be written for each condition of the truth table shown:

$A \times (B + C1) = C$ $1 \times (0 + 0) = 0$

$0 \times (0 + 0) = 0$ $1 \times (0 + 1) = 1$

$0 \times (0 + 1) = 0$ $1 \times (1 + 0) = 1$

$0 \times (1 + 0) = 0$ $1 \times (1 + 1) = 1$

$0 \times (1 + 1) = 0$

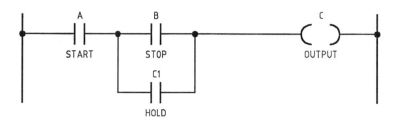

Figure 4.16   Boolean expression for a start/stop curcuit

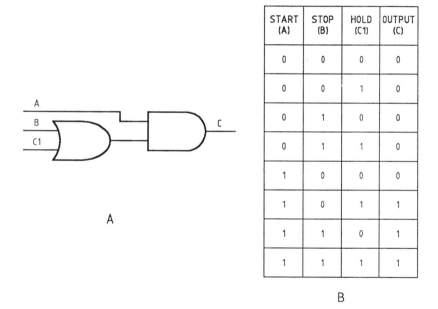

START (A)	STOP (B)	HOLD (C1)	OUTPUT (C)
0	0	0	0
0	0	1	0
0	1	0	0
0	1	1	0
1	0	0	0
1	0	1	1
1	1	0	1
1	1	1	1

B

Figure 4.17(a) Logic symbol equivalent to start/stop ladder diagram   (b) truth table

For any logic circuit a Boolean expression can be written and a resultant ladder diagram produced. Figure 4.18 shows a group of logic gates which produce the required logic for a programmable controller as shown in Figure 4.19. The array of gates consists of:

- gate 1, an AND function;
- gate 2, an OR function;
- gate 3, an AND function.

75

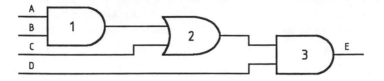

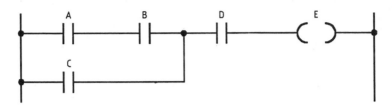

Figure 4.18   An array of logic gates

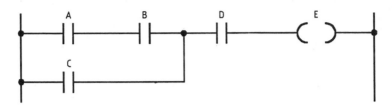

Figure 4.19   Ladder diagram solution for an array of logic gates

An output, or logic 1, at output E will occur when C and D are logical 1, or when A and B and D are logical 1, or when ABC and D are logical 1. A truth table and Boolean expression is then developed.

Each expression that produces the output can easily be determined by the Boolean expression. Although 2 cannot exist in Boolean algebra, it is understood that if a 2 appears in the answer the output would be active, but the output to produce the input may be invalid. Figure 4.19 is the ladder diagram that will produce the required output for the Boolean expression shown in Table 4.2.

Table 4.2   Boolean expressions developed from Figure 4.18

A B C D E	$A \times B + C \times D = E$
0 0 0 0 0	$0 \times 0 + 0 \times 0 = 0$
0 0 0 1 0	$0 \times 0 + 0 \times 1 = 0$
0 0 1 0 0	$0 \times 0 + 1 \times 0 = 0$
0 0 1 1 1	$0 \times 0 + 1 \times 1 = 1$
0 1 0 0 0	$0 \times 1 + 0 \times 0 = 0$
0 1 0 1 0	$0 \times 1 + 0 \times 1 = 0$
0 1 1 0 0	$0 \times 1 + 1 \times 0 = 0$
0 1 1 1 1	$0 \times 1 + 1 \times 1 = 1$
1 0 0 0 0	$1 \times 0 + 0 \times 0 = 0$
1 0 0 1 0	$1 \times 0 + 0 \times 1 = 0$
1 0 1 0 0	$1 \times 0 + 1 \times 0 = 0$
1 0 1 1 1	$1 \times 0 + 1 \times 1 = 1$
1 1 0 0 0	$1 \times 1 + 0 \times 0 = 0$
1 1 0 1 1	$1 \times 1 + 0 \times 1 = 1$
1 1 1 0 0	$1 \times 1 + 1 \times 0 = 0$
1 1 1 1 1	$1 \times 1 + 1 \times 1 = 2$

The Hitachi P 250E has OR, AND, and exclusive OR (EOR) functions available as part of the programmable controller's arithmetic set. The logic functions are word functions where the whole word is involved in the logic operation. Figure 4.20 is the ladder diagram and keystrokes required to program the OR function. The operation of this function is that when X 200 is true the logical OR function is carried out on word 15 and word 25 and the result is stored at word 35. As the word is made up of 16 bits of data, all 16 bits are operated on at one time. Figure 4.21 shows the result of the OR function where any time a 1 appears in word 15 or word 25, then a 1 will appear in the resultant word 35.

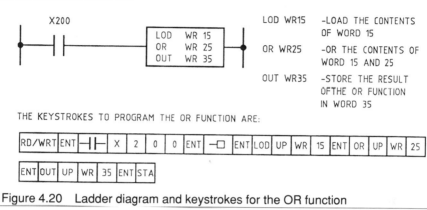

Figure 4.20   Ladder diagram and keystrokes for the OR function

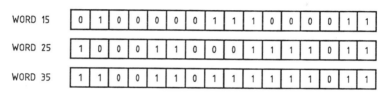

Figure 4.21   Result of OR function on word

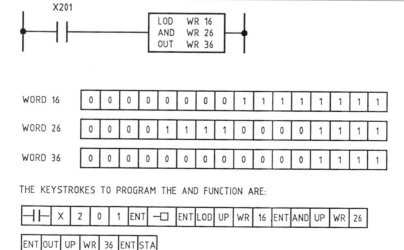

Figure 4.22   Programming the AND function

The keystrokes necessary to program the AND function, and the resultant operation on the associated words, is shown in Figure 4.22. The operation of this circuit is that when input X 201 is true an AND function of words 16 and 26 is carried out and the result located at word 36. To obtain a 1 in word 26 a 1 must be located at the same bit number of word 16 and word 26.

The keystrokes necessary to program the exclusive OR (EOR) function, and the resultant word operation, are shown in Figure 4.23. The operation of this circuit is that when X 202 is true the OR function is carried out on word 17 and 27 and the result located at word 37.

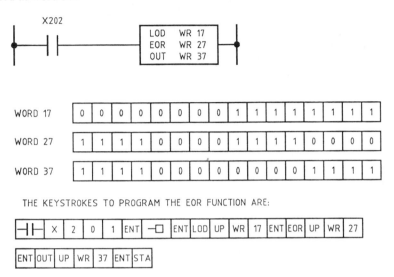

Figure 4.23   Programming the EOR function

4.7 Program jumps

Most programmable controllers have an operator command that allows the program to be carried out in a different order from the order in which it was inserted by the programmer. This type of instruction is a jump (see Figure 4.24). The jump instruction has a number of useful functions:

- It reduces scan time because unnecessary parts of the program do not have to be scanned.
- It allows the programmable controller to hold more than one program and scan only the program appropriate to operator requirements.
- It allows sections of a program to be jumped when a production fault occurs.

The jump instruction can have a conditional logic program preceding it in a rung so that the jump will only occur under predetermined conditions. Care must be taken when programming a jump or a number of problems may result:

- Jumping over counters and timers will stop them from being incremented.
- Jumping to locations that may not be allowable might cause the program scan to

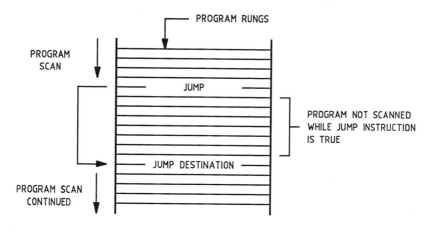

Figure 4.24   Jump operation

become lost. Jumping backward, jumping to an undefined destination or label, or jumping an end program instruction could cause this problem.

- Jumping to a destination within master or zone control areas could result in unpredictable machine operation because the program would be executed from the jump destination irrespective of the condition of the master or zone logic.

The Allen-Bradley programmable controller has a jump instruction which is identified by a two-digit octal number. When the condition area of the rung containing the jump instruction is true the program will jump to the destination of a label with the octal number corresponding to the jump instruction.

An example of a program jump is shown in Figure 4.25. When the contact 11102 in rung 2 is true the program will jump to label 02 in rung 6. The jump is entered in the output area of the rung using the JMP key and the label for the jump is entered in the condition area of the rung using the LBL key. Timer 031 is programmed in the jumped rungs of the program and will not be incremented. Timer 030 is programmed twice, once inside the jumped area and again outside the jumped area, to ensure the timer is incremented correctly.

The Hitachi P 250E allows jumps to be programmed into any program. Each jump has a reference number between 0 and 49. A jump is indicated by a JPS (jump start) instruction and must have an accompanying JPE (jump end) instruction after the JPS. Jumping backward in the program is not allowed. A jump is made in a program when the condition of a rung containing the jump is true. The jump will cause the rungs between the JPS and JPE not to be scanned, and if timers and counters are jumped they will not be incremented. It is considered bad programming practice when programming jumps to:

- jump in to a master control area as the program will continue to execute from the JPE and could cause haphazard operation of a machine or process or danger to personnel;
- jump counters and timers as they cease to increment when jumped.

79

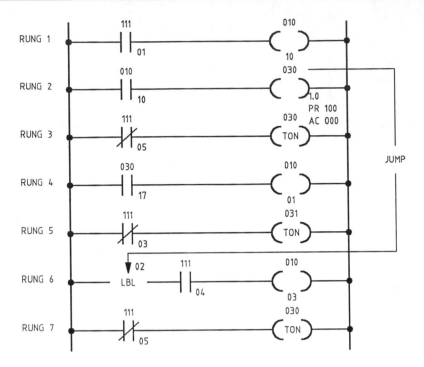

Figure 4.25   Programming jumps

Jumps cannot be made unless a JPE has been defined. Nested or overlapped jumps can cause JPS or JPE instructions to be missed.

Figure 4.26 is a jump instruction program for the Hitachi P 250E where, while rung 1 is true and passing power, the jump is in operation and timer TD 1 will not be incremented. TD 2, being outside the jump area, continues to be incremented.

Multiple jumps to a single label

The Allen-Bradley allows more than one jump to be made to a single label, thus allowing selected rungs of a program to be jumped and program functions to be grouped within a number of rungs. Figure 4.27 shows an example of groups of selected rungs that are jumped and where the same jump label is used. The condition area of the rung decides in which rung a jump will take place.

4.8 Program subroutines

A subroutine is a subprogram within a main program which is intended to carry out a specific function. It is often used when it is necessary to carry out a function a number of times but the program for the function is large enough to slow down the PLC processor, affecting the machine or process operation.

Subroutines are bounded by a jump-to-subroutine instruction, which causes the next rung to be the first rung in the subroutine, and a return command which causes

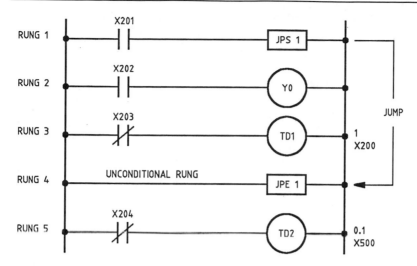

Figure 4.26   Hitachi P 250E jump instruction

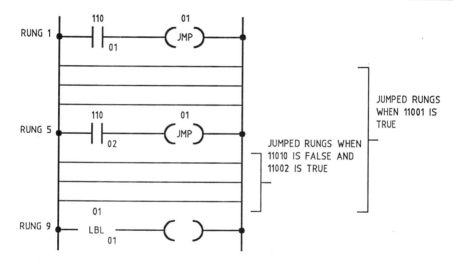

Figure 4.27   Multiple jumps to a single label

the next scanned rung in the program to be the one following the jump-to-subroutine instruction. A PLC program may contain several or many subroutines, or may have subroutines within subroutines which are called nested subroutines.

Allen-Bradley subroutines

Allen-Bradley PLCs have an area between the end of the main program and the message storage area set aside for subroutines. The start of the subroutine area is displayed on the industrial terminal and serves as the end-of-program statement for the main program. To establish the subroutine area the cursor on the industrial terminal is located at the end of the user program and the SHIFT and SBR key are pressed.

Sixty-four subroutines can be programmed into the Allen-Bradley PLC 2/17, each having its own unique identification label. To jump to a subroutine, a JSR output instruction is programmed with a two-digit octal address to indicate which subroutine is being called. A return (RET) instruction at the end of the subroutine causes the program to return to the rung containing the JSR plus 1, as shown in Figure 4.28. Multiple jumps to one subroutine are allowed.

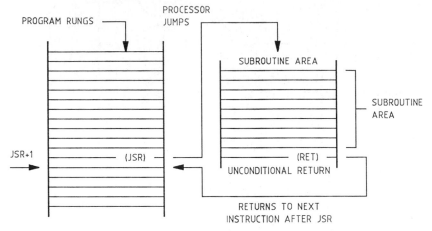

Figure 4.28 Jump to subroutine

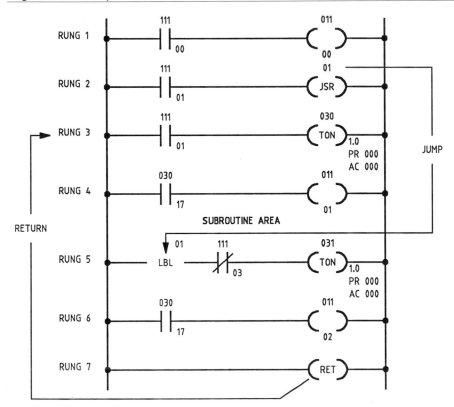

Figure 4.29 Subroutine ladder diagram example

In Figure 4.29 rung 2 will cause the processor to jump to the subroutine 01 in rung 5. Timer 031 will increment while the subroutine is being scanned. The return in rung 7 will cause the processor to jump back to the main program and start scanning from rung 3.

Nested subroutines

Some programmable controllers allow nested subroutines to be called from within a subroutine. Nested subroutines make complex programming easier, and program operation faster, as the programmer does not have to continually return from one subroutine to enter another. Programming nested subroutines may cause scan time problems because, while the subroutine is being scanned the main program is not, and excessive delays in scanning the main program may cause the outputs to operate later than required. Looping within a subroutine also causes excessive delays in scanning the main program. To overcome subroutine scan delay problems a scan counter can be used to count the number of times a subroutine is scanned between scans of the main program. Excess delays in scanning will cause a watchdog timer to produce a subroutine error display. Figure 4.30 represents nested subroutines.

The subroutine concept is the same for all programmable controllers, but the method used to call a subroutine and return from a subroutine uses different commands depending on the PLC manufacturer. The Hitachi P 250E uses CAL to call a subroutine, LB to label (start) a subroutine, and FUN 1203 to return to the rung after the CAL instruction which called the subroutine. The FUN instruction indicates that a program function is to be carried out. The CAL and LB instruction must have a corresponding identification between 0 and 99. Figure 4.31 shows the format of programming a subroutine into the Hitachi P 250E.

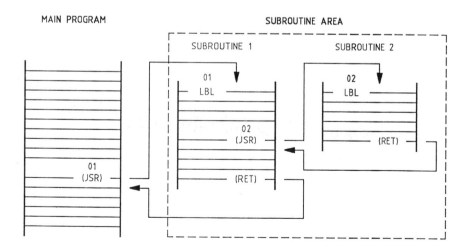

Figure 4.30   Nested subroutines

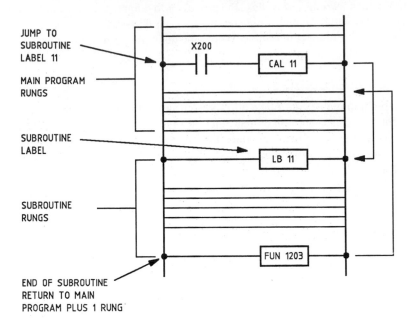

Figure 4.31   Hitachi P 250E subroutines

4.9 Test questions

Multiple choice

1 Which of the following are considered to be programmable controller modes of operation?
   (a) program.
   (b) on/off.
   (c) write.
   (d) serial/parallel.
   (e) edit.

2 A programmable controller used in a system instead of hardwired electronic or relay control will:
   (a) reduce documentation costs.
   (b) operate in an industrial environment.
   (c) not detect operational errors.
   (d) use discrete timers and counters.
   (e) interface with computers.

3 The type of input that is associated with machine or process control is:
   (a) analog input.
   (b) digital output.

(c) radio frequency input.

(d) noise input.

4 Which of the following is considered to be a peripheral associated with a programmable controller?

(a) Program terminal.

(b) Mass storage system.

(c) Analog input.

(d) Digital input.

5 The scan time for a programmable controller is the sum of:

(a) input and output scan time.

(b) user program scan time.

(c) user program and output scan time.

(d) user program, input and output scan time.

True or false

6 On line programming allows changes to be made to the user program while the PLC is continuing to operate.

7 When the programmable controller is in the run mode the programmable controller can be programmed.

8 The term multitasking when discussing programmable controllers is the term used to describe the condition of multiple inputs to the PLC.

9 A NOT function can also be termed an inverter.

10 A program jump is associated with a fault condition.

Written response

11 Draw and label a ladder diagram program for:

(a) an AND function.

(b) an OR function.

12 Draw and label a ladder diagram program containing a program jump.

13 List three precautions necessary when programming program jumps or jump to subroutines.

14 Write a Boolean expression for the logic gates shown in Figure 4.32.

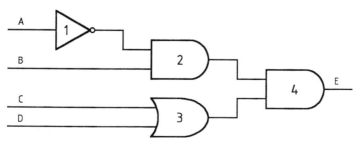

Figure 4.32   Question 14

15 Draw the truth table for a two-input exclusive OR function.

Timers and counters

5.1 Counting and timing functions in industrial control

The provision of timing and counting functions in programmable controllers was necessary to allow the controller to replace the function of hardwired electronic control that was required in machines or processes. The advantage of PLC counters and timers is that their specifications can be altered, or the number of them used in a circuit can be increased or decreased, by programming changes without wiring changes.

The number of counters and timers provided in a PLC will vary between manufacturers. There are techniques for increasing the number available, but, as the number of counters and timers is largely based on design criteria, the system designer must decide on the required number of counters and timers and I/O prior to purchasing the programmable controller.

Counters and timers in machine control and industry processes are used in almost every installation. Consider the following simple industry application where manufactured units on a production line are sorted and packaged according to size (see Figure 5.1). The system is initiated by an operator each day, and the items to be sorted are then fed onto the production line conveyor from a holding hopper. The items, once on the conveyor, proceed, operating limit switch 1 which counts all of the items. Limit switch 2 counts only the larger items. As limit switch 2 is operated a pneumatic ram is activated to cause all the larger items to be located in packing box 2. The smaller items continue to the end of the conveyor and are located in packing box 1. At the end of work, when the operator presses the end of work button, the conveyor must continue to run until all the items currently on the conveyor are cleared and packed. To produce the required logic for this system, conventional relay logic using electromechanical counters and an electronic timer could achieve the desired result.

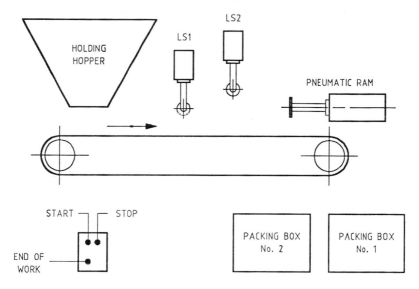

Figure 5.1   Conveyor sorting system layout

The counters count the large, as well as all the items, on the conveyor so that at the end of each day a figure for all the small and large items packed can be calculated. The timer is used to cause the conveyor to 'run on' after the end of work button has been pressed. Figure 5.2 is the electrical ladder diagram suitable to operate the conveyor system shown in Figure 5.1. The operation of the circuit in Figure 5.2 is that when power to the system is applied, the off-delay timer, rung 7, is energized via the end-of-work EW01 contact. The timer-enabled contact in rung 1 closes immediately and readies the system for a start.

The operator pressing the start button causes rung 1 to become true. The start relay energizes via the normally closed stop button, the start button which is now operated, and the timer contact which is now closed. The operation of the start relay provides a seal or latching circuit for itself via ST1 in rung 1. ST2 starts the conveyor motor, rung 2, and ST3 prepares the end-of-work relay circuit, rung 6.

As items proceed along the conveyor, limit switch 1 counts each item by causing rung 3 to become true as each item passes, thus energizing electromechanical counter 1. As a large item passes along the conveyor, limit switch 2 operates and causes the pneumatic ram to operate, placing them into packing box 2. Smaller items will pass under limit switch 2 and continue into packing box 1. The end-of-work button momentarily makes the circuit of the end-of-work relay in rung 6, which latches via its own contact. The timer period times for a predetermined duration which causes the conveyor to continue for the time necessary to sort all items currently on the conveyor. At the end of the timed period, rung 1 goes false, thereby releasing the start relay causing the system to stop.

The system can be stopped at any time by operating the stop push button in rung 1. To provide operator and machine safety an independent hardwired emergency stop facility must be included in the system. Any electrical, electronic, or programmable controller installations must have separately hardwired emergency stop facilities.

This example of electrical control, where counting and timing is undertaken using discrete components, can be replaced with a programmable controller using internal counters, timers, and relay logic. The same system will be described in terms of programmable controller operation in Section 5.9.

5.2 Counter and timer operation characteristics

A description and timing diagram of the most common counters and timers is provided in this section to assist the reader who is unfamiliar with timing and counting circuits to understand timer and counter operation.

The count function available in a programmable controller can be caused to count up to a preset value or count down to a preset value. The up-counter is incremented by 1 each time the rung containing the counter is energized. The down-counter

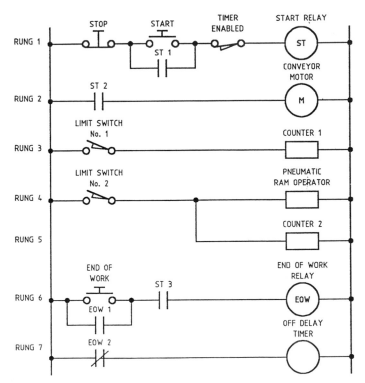

Figure 5.2   Conveyor sorting application

decrements by 1 each time the rung containing the counter is energized. Figure 5.3 shows the counting sequence of an up-counter and a down-counter. The value indicated by the counter is termed the accumulated value. The counter will increment or decrement depending on the type of counter until the accumulated value of the counter is equal to or greater than the preset value, at which time an output will be produced. A counter reset is always provided to cause the counter accumulated value to be reset to a predetermined value.

On electromechanical counters, the reset is usually manual; on programmable controller counters it may be under the control of the program. The preset value of a programmable controller counter can be programmed by the operator or can be loaded into the memory register as a result of a program decision.

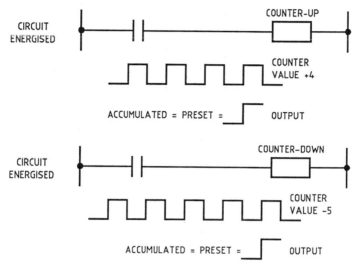

Figure 5.3   Counter operation sequence

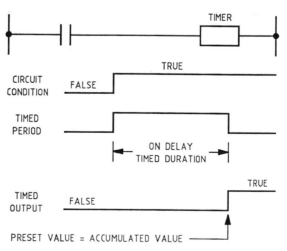

Figure 5.4   On delay timer

89

The on-delay timer operates such that when the rung containing the timer is true the timer time-out period commences. At the end of the timer time-out period an output is made active as shown in Figure 5.4. The timed output becomes active sometime after the timer becomes active, hence the timer is said to have an 'on' delay.

The timers available in programmable controllers will vary depending on the PLC manufacturer, but all will allow timers to be programmed in such a way as to produce on and off delays, and most will allow watchdog and single shot functions (discussed later in this chapter). The off-delay timer operation will keep the output energized for a timed period after the rung containing the timer has gone false. The timing diagram for the off-delay timer is shown in Figure 5.5.

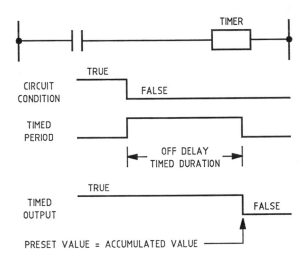

Figure 5.5   Off delay timer

Counter and timer addresses are usually specified by the programmable controller manufacturer and are located in a specific area of the data organization table. The Allen-Bradley PLC 2 family, as factory configured, has forty counters or timers starting at address 030 octal where the accumulated value is stored. The counter/timer preset values are located at 100 octal after the accumulated address at locations 130 octal to 177 octal. The Modicon Micro 84 has the preset value located at memory register addresses of 300X and 40XX and the accumulated value at address 40XX. The Hitachi P 250E has counters and timers located at addresses from 000 to 255; they are prefixed with the letters TD, SS, WDT, CU, and CL depending on the required function.

5.3 Counters

The counter instructions available on the Allen-Bradley PLC 2 series of programmable controller, Modicon Micro 84, and Hitachi P 250E are shown in Table 5.1.

To program a counter or timer in the Allen-Bradley system the following information about the counter/timer words is necessary. The accumulated word consists of sixteen bits. The first 12 bits of the word are the BCD value of the accumulated value.

The last four bits, 14, 15, 16, and 17 are used as the:

- increment bit (17);
- decrement bit (bit 16);
- output bit – accumulated value greater than or equal to preset value (bit 15);
- counter overflow/underflow bit (bit 14).

Table 5.1 Counter instructions

Allen-Bradley	Modicon Micro 84	Hitachi P 250E
CTU–up-counter	up-counter	CU–up-counter
CTD–down-counter	down-counter	CL–counter clear
CTR–counter reset	counter reset	

Figure 5.6 shows the accumulated and preset words for a counter and timer including the bits associated with timer and counter function.

Figure 5.7 shows the ladder diagram to program the Allen-Bradley PLC 2 family of counters. The operation of the ladder diagram is as follows:

1 When rung 1 is true counter 031 is incremented by 1. The accumulated value of the counter is displayed under the counter symbol on the ladder diagram. Bit 03117, rung 3, also becomes true each time the counter is enabled, causing output 01101 to become active.

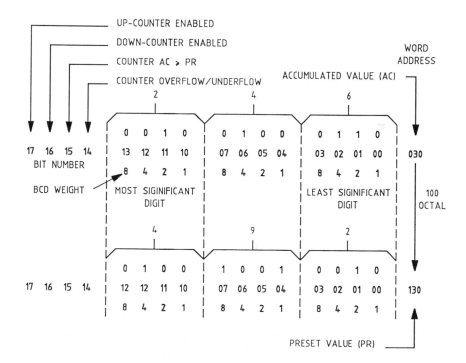

Figure 5.6   Allen-Bradley PLC 2 counter timer word/bit allocation

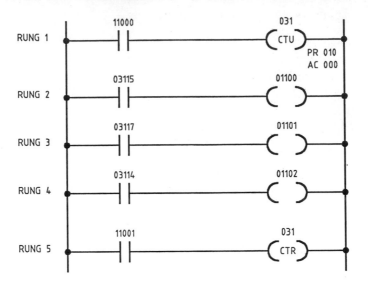

Figure 5.7   Allen-Bradley counters

2 The preset and incremented values, when equal, cause bit 03115 to be enabled or made active causing output 01100 to become active.

3 When the counter accumulated value is equal to or exceeds the preset value, bit 03115 is true.

4 Overflow has occurred when the count has exceeded 999. Bit 03014 becomes active causing rung 4 to become true and therefore bit 01102 active. The overflow bit is reset when the counter is reset.

5 To reset the counter's accumulated value a counter reset command is used. The counter reset command, rung 5 in Figure 5.7, will cause the accumulated value of the counter to be zero when input 11001 is active. The counter reset command must be addressed to the particular counter to be reset. If mains power failure occurs, the PLC counter accumulated value will remain the same. The processor will not respond to input signals, however, therefore losing counts during the power failure.

The Allen-Bradley down-counter has a similar operation to the up-counter, except that each time the rung containing the counter is made true the counter accumulated value is decremented by one. Bit 16 is true when the down counter is enabled. Bit 14 indicates when the counter is decremented past 000, i.e. underflow, and bit 15 is true when the preset and accumulated values are equal ,or the accumulated value is greater than the preset value. Resetting the down counter causes the preset value to have a value of 000 decrementing to 999 when next the counter is enabled, followed by 998, 997, 996, etc.

The Modicon Micro 84 also has a counter facility. This counter is incremented by one each time the rung logic containing the counter is true. The counter has a preset value of between 000 and 999 located in a register prefixed by 300X or 40XX. The accumulated value is located in a second register prefixed 40XX.

Figure 5.8 shows the ladder diagram of a counter where two outputs are connected to the counter. Output 1 will become true and remain true, until reset, when the counter

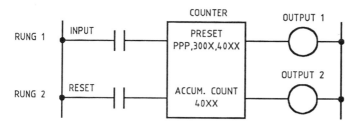

Figure 5.8   Modicon counters

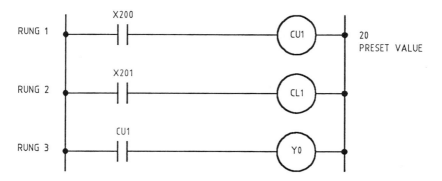

Figure 5.9   Hitachi P 250E counters

preset value equals the accumulated value. A second output will be true while the preset and accumulated values are not equal. To reset the counter an enable reset function is provided. When the enable reset is active the accumulated value of the counter is reset to zero.

The counter in the Hitachi P 250E uses two lines of logic as shown in Figure 5.9. The first line contains the counter coil, and the second line the counter reset logic. The operation of the ladder diagram is that when input X200 is closed the up-counter (CU1) is incremented by one. When the preset and accumulated values are equal, CU1 input is operated and causes rung 3 to become true, therefore causing output Y0 to be active. Y0 will remain active until the counter is reset. The maximum value that can be programmed as a preset value in the Hitachi P 250E is 4095. To reset the counter preset value to 000, rung 2 containing the counter reset CL1 is made active.

5.4 Timers

Table 5.2 shows the instructions available in the Allen-Bradley PLC 2 series of programmable controller as well as the Modicon Micro 84 and Hitachi P 250E.

Table 5.2   Timer instructions

Allen-Bradley	Modicon Micro 84	Hitachi P 250E
TON–on delay	on delay	TD–time delay
TOF–off delay	off delay	SS–single shot
RTO–retentive	retentive	WDT–watchdog

Allen-Bradley timers

Figure 5.10 shows a ladder diagram containing an Allen-Bradley on-delay timer. When input 11100 closes, rung 1 becomes true, causing timer 030 to commence its timed period. Timer 030 bit 17 closes causing rung 3 to become true, therefore output 01001 indicates the timer is enabled. The timer is incremented by a 1-second clock. The accumulated value is displayed at the timer symbol on the ladder diagram on the industrial terminal.

When the accumulated value and the preset value are equal, bit 03015 becomes active, therefore rung 2 on the ladder diagram is true and produces an output. The output produced is delayed for the duration of the timed period. The timed output will remain true until the timer is reset. To reset the timer-accumulated value to 000, rung 1 containing the timer must go false. Power failures cause the timer to stop timing, the accumulated value to be reset to zero, and the status bit to be reset.

The off-delay Allen-Bradley timer operates in a similar manner, except that when the rung containing the timer goes false, there is an off delay until the timed contacts go false. The off-delay timer uses bit 17 when the timer is enabled, and bit 15 as the timed bit.

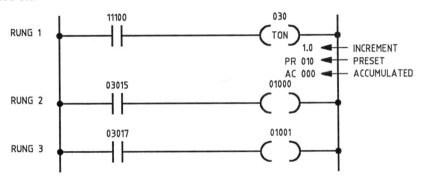

Figure 5.10   Allen-Bradley on delay timer

Retentive timer

The retentive timer retains the accumulated value of the timed period even if the input condition of the timer rung goes false. To reset this timer, a retentive timer reset command is used. Figure 5.11 shows a ladder diagram with a retentive timer and a retentive timer reset rung.

When rung 1 is true the timer will commence timing. If rung 1 goes false the timer will stop timing but will recommence timing from the stored accumulated value each time the rung goes true. Bit 03515 is the timed bit and bit 03517 is the timer enabled bit. To reset the accumulated value of the timer to 000, rung 4 must go true, causing the retentive timer reset to become active.

The timed rates for the Allen-Bradley timer are 1 second, 100 milliseconds and 10 milliseconds.

Figure 5.12 shows the ladder diagram for a Modicon timer. When the input on rung 1 is true the timer will commence to time. Output 2 of the timer will be active while the preset value is not equal to the accumulated value. When the preset and accumulated values are equal, output 1 will go true and output 2 false. The timer is reset when the enable-reset is made active. The Modicon Micro 84 timer has two timed rates, 1 second and 100 milliseconds. The timer contents are not lost during power failure, but the timer will cease to increment during the failure.

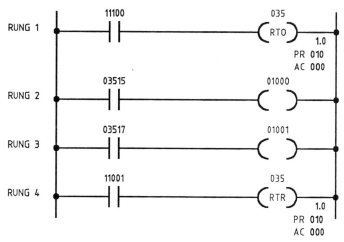

Figure 5.11   Retentive timer

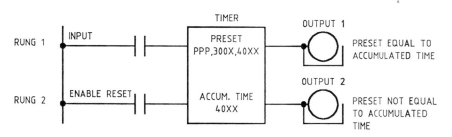

Figure 5.12   Modicon timer

The Modicon timer can produce four types of timing sequence depending on how the timer's inputs and outputs are configured. These configurations are:

- closing input on active output (normal output) – Figure 5.13(a);
- closing input off active output (not output) – Figure 5.13(b);
- opening input off active output (normal output) – Figure 5.13(c);
- opening input on active output (not output) – Figure 5.13(d).

Figure 5.13 shows the Modicon timer with the timing sequence and symbol for each of the four configurations. The Modicon timer can be programmed as a retentive type or non-retentive type depending on the application. The Modicon retentive and non-retentive timers are shown in Figure 5.14.

The Hitachi P 250E has three types of timer:

- an on-delay timer, where an output will become active after a delay period;

95

- a watchdog timer, to ensure that an event occurs between two time limits;
- a single shot timer, to provide a pulse which provides an output for the duration of the preset value of the timer.

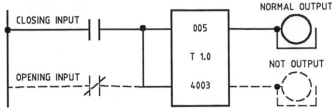

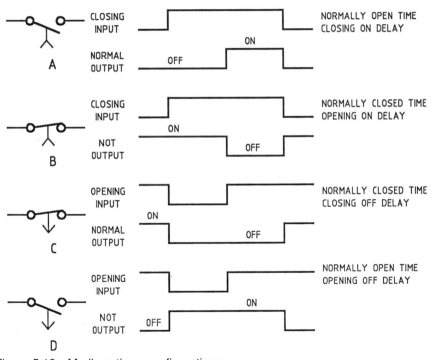

Figure 5.13   Modicon timer configurations

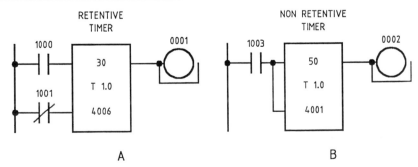

A   B

Figure 5.14(a) Modicon retentive timer   (b) non-retentive timer

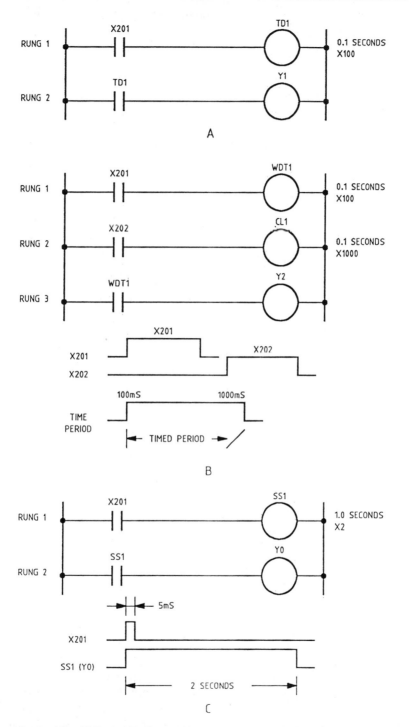

Figure 5.15 Hitachi P 250E timers (a) on delay timer (b) watchdog timer
(c) single shot timer

The Hitachi P 250E on-delay timer is abbreviated to TD. The timer output will energize after the time delay period, which will occur when the preset value equals the accumulated value. The preset value is a maximum of 12 bits which represents 4095 time increments which will accumulate at multiples of 1 second or 100 milliseconds. The preset value may be defined as a constant or as a word address under the program control. The program for a Hitachi on-delay timer is shown in Figure 5.15(a). Input X201, when closed, will cause the rung to be true and the timer will increment. When the input is false the timer will reset. When the accumulated value of the timer equals the preset value TD1 will operate, causing output Y1 to become active.

A watchdog timer ensures that an event occurs between two specified time limits. The watchdog timer is often used in a PLC system to detect timing errors or critical timed periods, as an output will occur if an input becomes true before a minimum time or if an input becomes true after a minimum time. In Figure 5.15(b), a watchdog timer, WDT1, is used to provide an output, Y2, if an event does not occur between 100 and 1000 milliseconds. X201 is the input that initiates watchdog timer operation. X202 must operate to clear the timer before 1000 milliseconds but not before 100 milliseconds output WDT1 produces an output if the time exceeds the preset timing limits.

The single shot or one shot timer provides an output, when enabled, for a duration equal to the preset value of the timer. The output produced is independent of the duration of the input signal. In Figure 5.15(c) input X201 is closed for 5 milliseconds. The timed period of the single shot timer SS1, however, produces an output Y0 for 2 seconds.

5.5 Cascading timers and counters

Counters and timers can be programmed in such a way as to cause an output of one timer or counter to increment the next. This method of programming counters and timers is termed cascading; it is used to provide counting or timed periods in excess of those provided by one counter or timer.

Allen-Bradley timers can be cascaded by using the timed bit (bit 15) of the first timer to enable the second. Each time the accumulated value of the first timer equals its preset value, bit 15 of the timer is set to a 1, thus the total accumulated value of both timers is equal to total accumulated time. The second timer in the cascade indicates the most significant digit and the first timer the least significant digit.

Allen-Bradley counters can be cascaded in a similar manner to the timer, but the overflow bit (bit 14) is used to increment the second counter. Figure 5.16 is an example of an Allen-Bradley cascade counter. The counter in rung 1 is incremented by input 11000. When the accumulated value of counter 030 exceeds 999 it energizes its overflow bit which in turn causes the second counter, 031, to increment by one. Counter 030 must be reset prior to continuing the count; this is achieved in rung 3 by using its overflow bit.

In the Gould Modicon Micro 84 timers and counters can operate in cascade by placing two in series. Timers connected in this manner allow the timed period to be cumulative. Figure 5.17 shows two Modicon timers in cascade, where the output produced by the first timer is the input for the second and reset for the first 0001, therefore causing the second timer to be a multiple of the first. An output 0002 is produced after each 100 clock pulses.

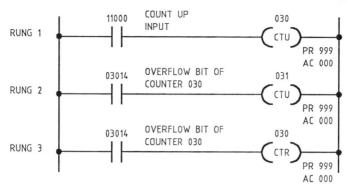

Figure 5.16 Allen-Bradley counters in cascade

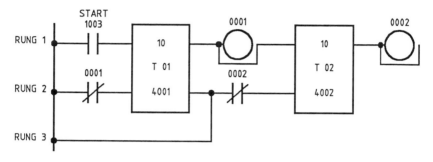

Figure 5.17 Modicon Micro 84 timers in cascade

The timers in the Hitachi P 250E can also operate in cascade where the output of the first timer becomes the input to a second timer. In Figure 5.18, timer 1 will increment for 4095 1-second increments, at which time timer 2 will be enabled by TD1. Timer 2 continues for a further 905 increments, at which time output Y0 will become active. Y0 opens timer 1 and 2 circuit and causes both timers to reset.

5.6 Up/down counter

The logic of an up/down counter can be obtained by programming an Allen-Bradley up counter in conjunction with a down counter using a common word address. An example of the up/down counter ladder diagram is shown in Figure 5.19, where the input on rung 1 will increment the counter-up and the input on rung 2 will decrement the counter-down at the same address. A reset of the counter will occur when the accumulated value equals the preset value.

5.7 Timers operating as oscillators

In some applications it is necessary to operate a timer as an astable multivibrator, or to produce a system clock with an even mark space ratio. To achieve the desired output two timers can be used as shown in Figure 5.20, where, in rung 1, timer TD 1 will time as soon as the PLC is switched to the run mode. After 20 seconds, the output from TD1,

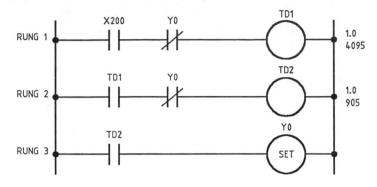

OUTPUT 'Y0' OCCURS AFTER 5000 SECONDS

Figure 5.18   Hitachi P 250E times in cascade

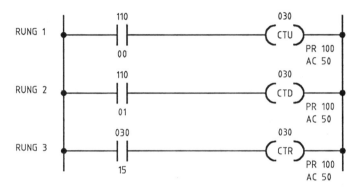

Figure 5.19   Up/down counter

as shown in rung 2, will cause output Y0 to be true, in addition, Y0 will cause rung 3 to become true and start timer TD2. At the timed period of TD2, rung 1 ceases to have power flow due to TD2 becoming not true which releases Y0, rung 2, and TD2 rung 3. The timing period then recommences. Output Y0 will have a 20-second mark/ space ratio as shown in the timing diagram in Figure 5.20, although the mark to space ratio of the output can be adjusted because it is dependent on the timer's timed period.

5.8 Timers and scan time

The timers in a programmable controller are incremented by an internally generated clock which originates in the processor module. The timer timebases are also generated from this clock. The majority of programmable controllers have timebases of 1 second and 0.1 second; some have a timer time base of 0.01 second. As PLCs will be accurate to within one internal clock pulse of the time base, the 0.01 second (10 millisecond) timer is valuable when the PLC is controlling high-speed events or it is necessary to generate short duration pulses. The 10-millisecond timer may cause problems when operating in conjunction with extremely long user programs.

100

Consider a typical PLC that has a scan cycle that scans input devices, user program, and output devices. As the scanning process is sequential and one instruction is scanned at a time, the scan time for the program is equal to input scan time plus output scan time plus program scan time. Scan time of the user program can vary from 15 milliseconds for a short program, to 50 milliseconds for a longer program. A typical scan time would be approximately 25 milliseconds. If the scan time is 25 milliseconds and there are 10-millisecond timers in the user program, the quickest that the timers can be incremented is every 25 milliseconds, therefore the 10-millisecond timers will lose time and become inaccurate. The method used to overcome this problem is to program the 10-millisecond timers or other devices that could be affected by long scan time into the user program more than once. The additional rungs will ensure that devices are scanned by the processor in a time less than the increment time of the device.

5.9 Timer and counter applications

Consider again the simple industrial application demonstrated earlier in this chapter, where items on a conveyor were being sorted according to size, counted, and a run-down time was required to clear the conveyor each day. Figure 5.21 is a ladder diagram of how this operation could be carried out by a PLC. The operation of the ladder diagram logic is:

Rung 1	11001	stop push button
	11002	start push button
	03015	off delay timer (normally closed)
	02000	storage bit (start)
	02000	start seal in
Rung 2	02000	start conveyor motor
	01100	conveyor motor contactor
Rung 3	11003	limit switch 1 count all components
	040	counter for all components
Rung 4	11004	limit switch 2 count small components
	041	counter for small components
Rung 5	11004	operate pneumatic ram
	01101	pneumatic ram operate solenoid
Rung 6	11005	end of work press button normally closed
	02001	end of work seal in
	02000	end of work start ready
	02001	end of work
Rung 7	02001	end of work contact
	030	off delay timer to provide conveyor run on

PLC traffic control signal system

In the following example a PLC is used to control a set of traffic signals at a crossroad. The roads run north/south and east/west. The light sequence is as follows:

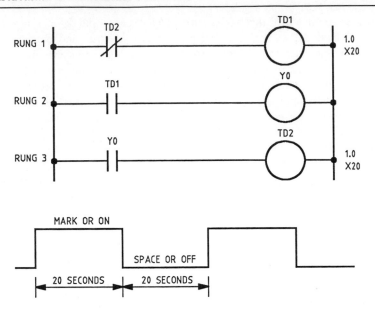

Figure 5.20   Timers operating as oscillators

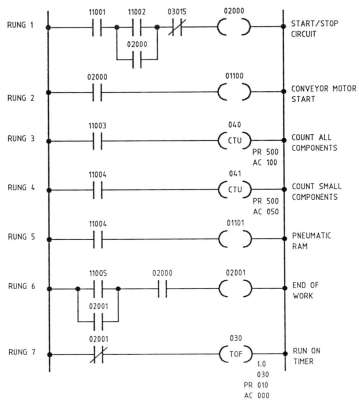

Figure 5.21   Conveyor sorting application PLC ladder diagram

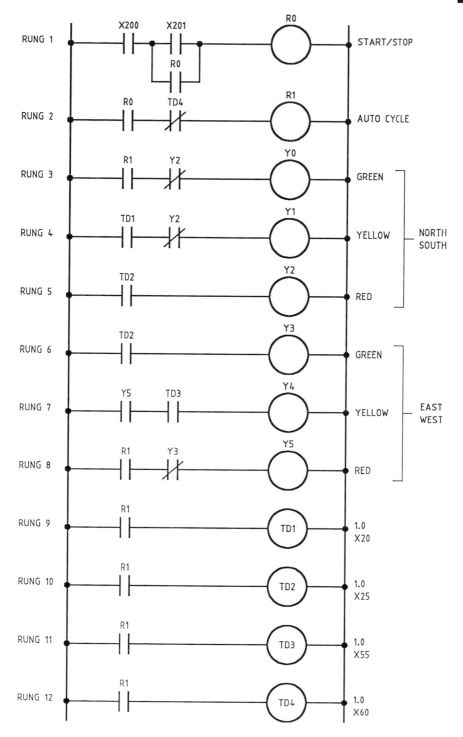

Figure 5.22   Hitachi P250 E PLC traffic signal control ladder diagram

start	north/south	Green on, red and yellow off
	east/west	Red on, green and yellow off
20 seconds later	north/south	Green and yellow on, red off
	east/west	Red on, yellow and green off
5 seconds later	north/south	Red on, green and yellow off
	east/west	Green on, red and yellow off
30 seconds later	north/south	Red on, green and yellow off
	east/west	Green and yellow on, red off
5 seconds later	north/south	Green on, red and yellow off
start of new cycle	east/west	Red on, green and yellow off

The output addresses for north/south are:

Red	Y 2
Yellow	Y 1
Green	Y 0

The output addresses for east/west are:

Red	Y 5
Yellow	Y 4
Green	Y 3

The ladder diagram program for this example (see Figure 5.22) is for the Hitachi P 250E, but it would work equally as well on any PLC.

5.10 Test questions

Multiple choice

1 The function of a watchdog timer is to:
   (a) time a period for an off delay.
   (b) time a period for an on delay.
   (c) time a period between a minimum and maximum.
   (d) produce an output for a preset time, independent of the input signal.

2 If mains power fails in an Allen-Bradley PLC 2
   (a) both preset and accumulated value in a counter will reset to zero.
   (b) preset values in a counter will reset to zero, accumulated will remain the same.
   (c) accumulated value in a counter will reset to zero and preset will remain the same.
   (d) the value in a counter will remain the same, but the counter will not increment.

3 When the rung containing a retentive timer output goes false, the values in the timer will:
   (a) reset to zero.
   (b) increment by 1.
   (c) remain the same.
   (d) decrement by 1.

4 A single shot timer:
  (a) works once only.
  (b) produces an output only once.
  (c) produces an output pulse for a predetermined time.
  (d) can be used as an up/down counter.
5 The scan time of a PLC is equal to:
  (a) the time taken to scan the user program.
  (b) the time taken to scan the inputs and outputs.
  (c) the speed which field contacts close and outputs can be operated.
  (d) the time taken to scan inputs, outputs and user program.

True or false

6 Timers are incremented in 1-second, 100-millisecond and 10-millisecond steps by the processor clock.
7 Cascade counters are connected in parallel to obtain counts in excess of 999.
8 Scan time must be considered when programming 10-millisecond timers.
9 The on-delay timer provides an output after a delayed period equal to the preset value of the timer.
10 Timers and counters can be located anywhere on the data organization table.

Written response

11 Sketch the ladder diagram for an up/down counter.
12 Sketch the ladder diagram for an astable multivibrator with a mark of 300 milliseconds and a space of 700 milliseconds.
13 Design a PLC ladder diagram for a department store sign that will sequentially turn on three lights 2 seconds apart then blink all three lights twice at a 2-second rate and repeat the sequence.
14 Produce a ladder diagram which will cascade timers to produce a 24-hour clock.
15 Design a ladder diagram that will count to 5000 then automatically reset.

Arithmetic functions

6.1 Introduction to word manipulation

In complex industrial machine or process control, it is necessary to carry out mathematical functions ranging from basic arithmetic, such as add, subtract, divide, and multiply, to complex mathematics used in flow measurement, or proportional integral and derivative control used in position control. In this chapter the basic arithmetic functions are examined in conjunction with the compare functions and the basic data manipulation functions necessary to carry out the arithmetic functions.

The arithmetic functions require the manipulation of multiple bits of consecutive data which are termed data words; programming examples in the previous chapters were oriented toward the manipulation of data bits. The advantage of a programmable controller being able to handle data in the form of 8-bit bytes or 16-bit words is that it allows analog values to be input to output from the PLC as well as multipoint input digital devices such as thumbwheel switches.

Input from thumbwheel switch

Thumbwheel switches are a type of digital input that require a number of consecutive bits to input BCD values. There are two types of thumbwheel switch: BCD, and binary.

A BCD thumbwheel switch is shown in Figure 6.1(a). Here a 4-digit BCD value is wired directly into the PLC so that word 115 is presented to the input terminals of a programmable controller as four bits of BCD per digit. As four bits are necessary

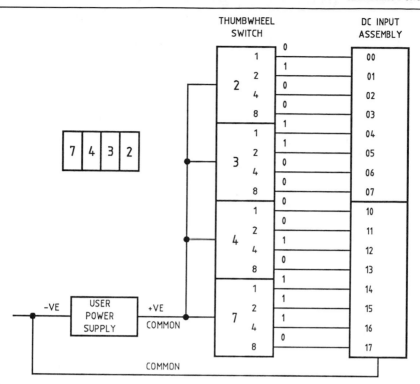

Figure 6.1(a)   Thumbwheel switch connection to the input assembly

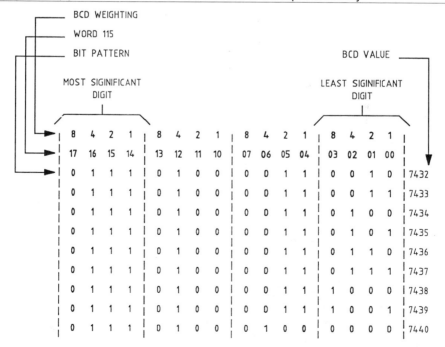

Figure 6.1(b)   Bit representation of BCD

to represent each BCD digit, 16 data bits are required to input the 4-digit BCD value. The length of the programmable controller word, that is, the number of bits, is determined by the type of microprocessor upon which the PLC is based. The length of the word governs the number of BCD digits that can be input.

Figure 6.1(b) shows a representation of the operation of the thumbwheel switch and how the switch affects the BCD digital inputs. The thumbwheel switch is constructed so that 1s appear in the output of the switch which represent the BCD equivalent of the decimal number displayed on the thumbwheel. As the thumbwheel switch is incremented, the bit pattern at its output increments. At 7439 (in Figure 6.1(b)) the next most significant digit must increment and the least significant is returned to zero.

There are three types of input derived from thumbwheel switches:

- decade, 1 pole 10 position;
- BCD, 1 pole 10 position;
- hexadecimal, 1 pole 16 position.

The type of switch most often used as a digital input for a programmable controller is a BCD type. The BCD output can be a 1 or high, which is approximately equal to supply voltage as the active state, or the complement type where inverted logic is used and 0 is the active state. Figure 6.2 shows the physical construction of a BCD thumbwheel switch.

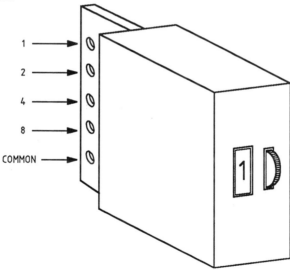

Figure 6.2   Thumbwheel switch

Seven-segment displays

Seven-segment displays are used to display numerical values of the BCD contents of registers within the PLC system. The seven-segment display can be liquid crystal (LCD) or light-emitting diode (LED); it can be driven directly from the PLC via a TTL output assembly, or via a DC output assembly, providing the display is interfaced to the DC assembly correctly. The seven-segment display has seven sections arranged in such a way as to produce an alpha numeric character which allows the display to produce hexadecimal outputs from A to F as well as decimal values between 0 and 9. The display segments are arranged as shown in Figure 6.3.

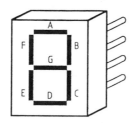

Figure 6.3   Seven segment display

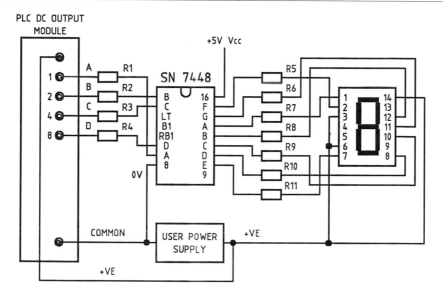

DECIMAL	LT	RB1	D	C	B	A	B1	A	B	C	D	E	F	G
0	1	1	0	0	0	0	1	1	1	1	1	1	1	0
1	1	X	0	0	0	1	1	0	1	1	0	0	0	0
2	1	X	0	0	1	0	1	1	1	0	1	1	0	1
3	1	X	0	0	1	1	1	1	1	1	1	0	0	1
4	1	X.	0	1	0	0	1	0	1	1	0	0	1	1
5	1	X	0	1	0	1	1	1	0	1	1	0	1	1
6	1	X	0	1	1	0	1	0	0	1	1	1	1	1
7	1	X	0	1	1	1	1	1	1	1	0	0	0	0
8	1	X	1	0	0	0	1	1	1	1	1	1	1	1
9	1	X	1	0	0	1	1	1	1	1	0	0	1	1

X = EITHER 1 OR 0

TRUTH TABLE

Figure 6.4   BCD to seven-segment decoder driver

Seven-segment LED displays can be either common anode or common cathode configured and are therefore purchased to suit the application.

The PLC cannot directly drive the seven-segment display with a BCD output; a BCD-to-segment decoder driver integrated circuit is required to produce the required segment patterns on the seven-segment display. The interface circuit required to drive a seven-segment LED display is shown in Figure 6.4.

GET and PUT instructions

To carry out basic arithmetic functions in the Allen-Bradley PLC, a GET instruction is used to read the contents of a register of a specific word address, and, on completion of the arithmetic function, a PUT instruction is used to put the information or answer into a register at another word address. Using the GET and PUT instruction to move data in words around the PLC memory is a form of data manipulation.

Figure 6.5 is a simple program that GETs the contents of a register at word 115, which in this case is a BCD thumbwheel switch, and PUTs the contents of register word address 115 into a register at word address 013 which has a seven-segment display wired to the output assembly terminals. If the PLC is in run mode, then, as the thumbwheel switch is operated, the seven-segment display will change to read the value indicated by the thumbwheel switch. The rung in Figure 6.5 is therefore unconditional; the output will transfer the data from the thumbwheel input to the seven-segment display each scan cycle.

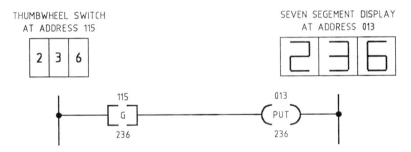

Figure 6.5   Allen-Bradley GET and PUT instruction

The GET instruction can also be used to transfer a value anywhere in the data organization table that is user accessible. Transferring information in this manner allows counter and timer preset and accumulated values to be altered at the operator's discretion.

A timer can have a value entered into the preset or accumulated data table location, or it could be reset to zero, if necessary, by providing the ladder diagram logic to shift the GET value as shown in Figure 6.6(a). In this application a timer 030 preset value is loaded from a thumbwheel switch. The thumbwheel is at word address 115 with an input value on the thumbwheel of 236. The thumbwheel value is PUT into address 130, which is the location of the preset value of timer 030, i.e. 100 octal down the data organization table when contact 11101 is closed and the rung true. In Figure 6.6(b) the accumulated value of timer 035 is transferred to a seven-segment display, but only when rung 3 is true via 11103 in the condition area of the rung.

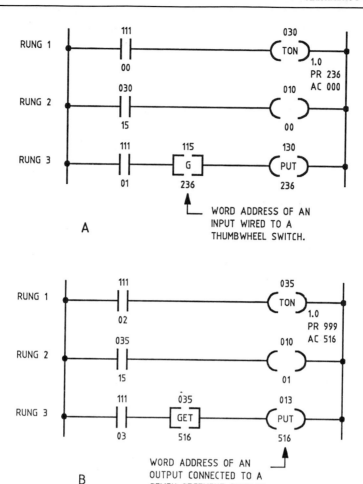

Figure 6.6   GET and PUT applications

6.2 Compare functions

Allen-Bradley data comparison instructions

Data comparison instructions do not shift values into or out of data table registers, but values at different register locations are compared and outputs made active or inactive depending on the results of the comparisons.

The data comparison instructions available on Allen-Bradley PLC 2 series are:

- EQUAL TO (=);
- LESS THAN (<);
- GET BYTE/LIMIT TEST.

The order that combinations of the EQUAL TO and LESS THAN instructions are programmed into the programmable controller memory allows the following additional instructions to be developed:

- GREATER THAN (>);
- GREATER THAN OR EQUAL TO (≥);
- LESS THAN OR EQUAL TO (≤).

The values for LESS THAN and EQUAL TO are either 12-bit BCD or octal values, whereas GET BYTE and LIMIT TEST operate on 8 bits or 1 byte of the compared data.

EQUAL TO and LESS THAN instructions

Figure 6.7(a) shows the EQUAL TO instruction and how it can be used to manipulate the output of timer 030 to latch an output derived from a timer for a portion of its timed period. The operation of this ladder diagram is as follows:

1 Rung 1 is true causing timer 030 to commence timing via its own contact 03015 which is used to set the timer.
2 The GET instruction is continually reading the accumulated value of the timer and comparing the GET value to a preset value at word locations 020 and 021 associated with the EQUAL TO instruction at rung 2 and 3.
3 When the GET value is equal to 015, rung 2 will become true, causing output 01100 to become active and latch.
4 Five timer increments later, when the GET value of the timer is equal to 025, output 01100 will be unlatched, rung 3. The timer will continue to time until it is reset by its own contact at a timed period of 30 seconds.

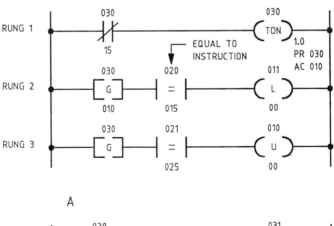

A

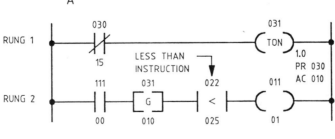

B

Figure 6.7(a) EQUAL TO instruction   (b) LESS THAN instruction

The latch is required so that the output will remain on for ten timer increments because the EQUAL TO instruction is only true while the GET value and EQUAL TO value are identical.

The LESS THAN instruction shown in Figure 6.7(b), rung 2, will compare the accumulated value 030 to the LESS THAN fixed value of 025 at address 022. While the timer increment is less than 025 and condition input 11100 is closed, the rung will pass power causing output 01001 to be active. When the timer increments to 025 the rung will cease to pass power and output 01001 will go false. The value of 025 at address 022 is fixed and is the reference value for the LESS THAN instruction.

The operation of the ladder diagram in Figure 6.8 is as follows:

Rung 1 is a self-resetting timer which will increment in 1-second pulses up to 990 prior to being reset.

Rung 2 is a GREATER THAN instruction, even though the LESS THAN function is used to generate the instruction. The GREATER THAN is obtained by placing the reference value 555 into the GET instruction and the LESS THAN value at the timer's address. While the accumulated value of the timer is below 555 there is no power flow through the rung and output 01102 is false. When the accumulated value of the timer exceeds 555, the rung has power flow and output 01102 is true.

Rungs 3 and 4 are the GREATER THAN, EQUAL TO instruction. The LESS THAN instruction is programmed to increment with timer 032 with an EQUAL TO instruction in parallel with the LESS THAN instruction. The reference value of 333 is located at address 024. While the accumulated value of timer 032 is less than 333 the rung is false and output 01103 is off. When the timer increments to 333, the

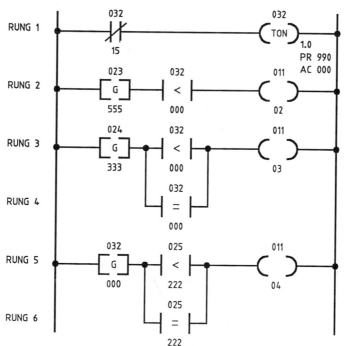

Figure 6.8 Greater than, greater than and equal to, less than and equal to instructions for Allen-Bradley PLC

EQUAL TO instruction becomes true, power flows through the rung, and 01103 turns on. When the timer increments to 334 the EQUAL TO instruction will go false, the LESS THAN will become true, and the power flow will continue through the rung to keep output 01103 on. Output 01103 will be true only when the GET value is greater than or equal to 333.

Rungs 5 and 6 produce the LESS THAN, EQUAL TO instruction. In this rung the reference value is located in the LESS THAN and EQUAL TO instruction at address 025; the GET at address 032 is incremented by the timer. While the timer-accumulated value is less than and equal to 222 the rung is true and output 01104 is on. When the accumulated value of the timer reaches 222 the LESS THAN instruction will go false, but output 01104 will remain on due to the EQUAL TO being active for increment 222. At timer increment 223 the rung goes false and output 01104 is off.

GET BYTE and LIMIT TEST

Allen-Bradley uses a GET BYTE and LIMIT TEST instruction together to compare an octal value to either the upper or lower byte of a 16-bit word. The GET BYTE and LIMIT TEST are programmed in the rung condition area, and, when true, allow power flow through the rung. The GET BYTE instruction will address the upper byte of a word if a 1 is programmed into the instruction, and will address the lower byte if a 0 is programmed into the instruction.

The LIMIT TEST instruction is addressed to a data table word which holds the reference values for the upper and lower limits to be tested. The upper test limit is located in the upper byte and the lower test limit in the lower byte of the address word.

Figure 6.9 is a ladder diagram rung which contains a GET BYTE and LIMIT TEST instruction programmed in such a way as to act as a limit switch. When the value of the lower byte of word 115 is greater than or equal to 111, or less than or equal to 222, power flow through the rung will take place and output 01110 will be true and on.

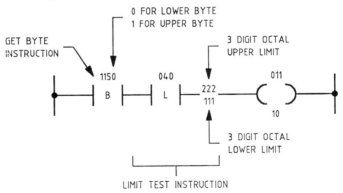

Figure 6.9   GET BYTE LIMIT TEST programmed as a limit switch

Hitachi P 250E arithmetic and data comparison instructions

The Hitachi P 250E has a range of arithmetic functions and compare functions similar to other PLC manufacturers. The terms used to describe the functions, however, can more often be associated with the terms used when discussing microprocessors.

TO PROGRAM THIS ARITHMETIC RUNG THE FOLLOWING
KEYSTROKES ARE REQUIRED.

WRT	⊣⊢	X	2	0	0	⊣☐
UP	LOD	UP	WR	1		ENT
UP	OUT	WY	2		ENT	STA

Figure 6.10   Load and OUT instructions

The Hitachi P 250E uses an arithmetic box to carry out 16-bit word manipulations to achieve the desired result. The rung containing the arithmetic box must be true to produce a result, and may be conditional or unconditional. The operation of each input instruction in Figure 6.10 is as follows:

1 WRT – places PLC into the write mode;
2 ⊣⊢– X200 – input with an address of 200;
3 ⊣☐ – arithmetic box;
4 UP – described as the shift key on some PLCs;
5 LOD WR10 – loads the contents at word 10;
6 OUT WR15 – outputs the data contained in word 15;
7 ENT – enters the information;
8 STA – stores the information in memory.

The operation of the ladder diagram in Figure 6.10 is that when input X200 is true the contents of word 1 are transferred to word 2; as the contents of word 1 vary so do the contents of word 2. When X200 is false the contents of word 2 are retained and not affected by changes to the contents of word 1.

Operation of the arithmetic register

Three registers, as shown in Figure 6.11, may be involved in an arithmetic function. These registers are:

• arithmetic register (AR) which is a 16-bit register;
• extended or expansion (ER) register which is a 16-bit register;
• carry register (C) which is a 1-bit register.

Word 1 is loaded into the arithmetic register and transferred to the extended register when the contents of the arithmetic register is output to word Y2.

To use the 1-bit carry register, the test bit (TS), and carry out (OUC) instructions are used. If bit 12 is set to 1 in word 1, a carry is generated. Using the test bit (TS) instruction on bit 12, a set or not set condition can be detected, and an output is placed into internal word R10 to indicate when a carry is generated.

To load a constant into an arithmetic register and output the contents of the arithmetic register to word address WY0, the ladder diagram rung and keystrokes shown in Figure 6.12 are required.

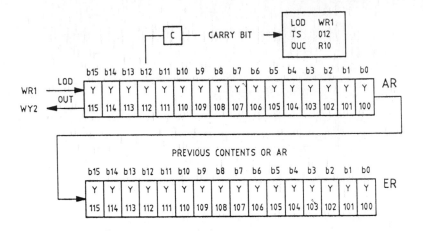

6.11 Hitachi P 250E arithmetic register operation

```
TO PROGRAM THIS ARITHMETIC RUNG THE FOLLOWING
KEYSTROKES ARE REQUIRED.
```

WRT	⊣├	X	2	0	1	─☐	LOD
UP	CON	1	2	3	ENT	UP	
OUT	UP	WY	0	ENT	STA		

```
CON - CONSTANT 123
WY 0 - WORD Y 0
```

Figure 6.12   Loading and constants instruction

Compare functions

The Hitachi P 250E supports three compare functions:

- CPH compare for greater than (higher than);
- CPE compare for equal to;
- CPL compare for less than.

The operation of the ladder diagram in Figure 6.13 is that X200, rung 1, causes the on-delay timer to increment in 1-second steps up to 20. Rung 2, when X201 is true, compares the word loaded from timer 1 with a constant 10; while the timer incremented value is equal to or greater than 10, output Y5 will be true and on.

In rung 3, while the timer increment is less than or equal to the constant 15, output Y6 will be true and on.

In rung 4, output Y7 will be true and on when the incremented value of the timer is equal to the constant 5.

116

6.3 Add, subtract, divide and multiply

The Allen-Bradley, Gould Modicon and most other PLCs have the ability to carry out add, subtract, divide, and multiply math functions. Some programmable controllers also have the ability to carry out the complex math functions used in process control.

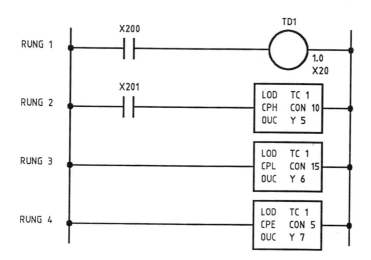

6.13 Hitachi P 250E compare functions

Allen-Bradley PLC 2/15

The Allen-Bradley PLC 2/15 can be programmed to add, subtract, divide, and multiply two 3-digit BCD values. The BCD values to be manipulated are stored in GET words in the condition area of a rung, and the arithmetic function to complete the operation is located in the output area of a rung. It is possible to program condition logic preceding an arithmetic operation. Condition logic in the rung must be true for the arithmetic operation to be carried out.

An arithmetic result is stored in the lower 12 bits of a data table word for add and subtract, and two data table words for divide and multiply. Bits 14 and 16 are used to indicate overflow and underflow respectively. The overflow bit is set to 1 when the arithmetic result exceeds 999. The underflow bit is set when the arithmetic result is a negative number.

Figure 6.14 shows add, subtract, divide, and multiply program rungs suitable for an Allen-Bradley PLC2. Rung 1 contains a counter which increments each time input 11000 is closed. The counter is reset by input 11001 in rung 6.

Rung 2 contains two GET instructions, the first being the accumulated value of counter 030, and the second the value from address 020. The value of the two GET instructions are added and the result stored at address 024. As the result exceeds 999, the overflow bit is set.

Rung 3 contains two GET instructions followed by the subtract instruction. The second GET instruction is subtracted from the first and the result is stored at location

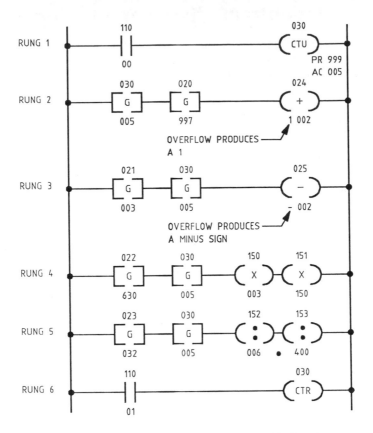

Figure 6.14   Allen-Bradley add, subtract, divide and multiply functions

025. As the second GET value is greater than the first, an underflow is indicated by a minus sign preceding the answer.

Rung 4 contains the multiply instruction preceded by two GET instructions. The multiply instruction consists of two consecutive words, making programming easier. The BCD value of the two GETs, when multiplied, are stored at address 150 for the most significant digits, and 151 for the less significant digits. Rung 5 is the divide instruction where the BCD value of the second GET in the rung is divided into the BCD value of the first GET. The result is stored in two consecutive addresses of 152 and 153. Address 152 holds the whole numbers and 153 the decimal fractions; a decimal point is located between the two addresses.

Hitachi P 250E

An arithmetic box is used to program the Hitachi P 250E to add two numbers. The boxes used on the Hitachi P 250E must have any logic condition true which precedes the box for the function within the box to operate. Rungs containing arithmetic boxes may be unconditional. An arithmetic box may contain up to nineteen instructions. Figure 6.15 is a ladder diagram containing an instruction which will add the contents of a word to a constant, and, if the result is greater than 65535, a carry bit will be generated and cause output Y0 to become active.

Figure 6.15 also contains an arithmetic box with a subtract function. A carry bit will be set if the result is negative. Word 1 is loaded into the arithmetic register, has word 3 subtracted from it, and the result is stored in word 4. If a negative result is produced, output Y1 will become active.

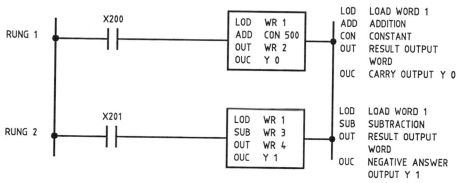

Figure 6.15   Hitachi P 250E add and subtract function

6.4 Using complex arithmetic

In many controlled processes it is necessary to carry out complex math functions to produce the desired output. Consider an application where a programmable controller is used to stream-switch four regulator runs in a metering station for a natural gas pipeline, as well as calculate the total gas flow. The values that are required for display and to calculate the flow are:

- static pressure;
- differential pressure across an orifice plate;
- temperature.

These values are analog values fed onto the programmable controller as 4 to 20 milliamp current loops; they are converted to their digital equivalent by a PLC analog input assembly.

Figure 6.16 shows the field arrangement of a metering station. To control the stream switching, a valve actuator is used with a fully-open and fully-closed limit switch wired to the PLC. The function of each section of Figure 6.16 is:

1 The regulator reduces the inlet pressure so that the downstream pressure is constant and relatively independent of flow.
2 The orifice plate is a flat plate with a hole in its center so that a differential pressure, which is dependent on gas flow, will be developed across the plate. The greater the flow of gas the greater the differential pressure. There is one orifice plate in each stream.
3 Valve actuators switch on the additional streams when demand is heavy, i.e. when flow is increased.

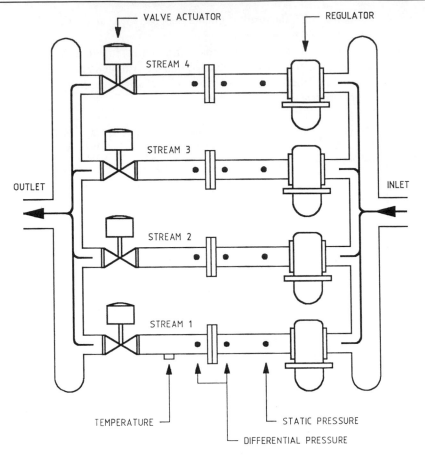

Figure 6.16   Stream switching and flow measurement application

4 Static pressure is a displayed value of the pressure upstream of the regulators.

5 Temperature is required for the flow calculation correction; it is measured downstream of regulators and orifice plate.

Static pressure and temperature are measured in stream 1 and assumed to be the same in each stream. Stream 1 would always be on even at the lowest flow conditions. The fundamental formula required to calculate flow is:

$$Q = EAo\sqrt{2gh}$$ where Q = flowrate (volume per unit time)
E = efficiency factor
Ao = area of orifice plate in square feet
g = acceleration due to gravity
h = differential pressure across the orifice in feet

This formula, suitable for an Allen-Bradley PLC 2/17 using the math functions available with the execute auxiliary function instruction, can be programmed into a programmable controller by using the ladder diagram shown in Figure 6.17. There are, however, a number of PLCs with math functions that could adequately carry out the flow calculation.

120

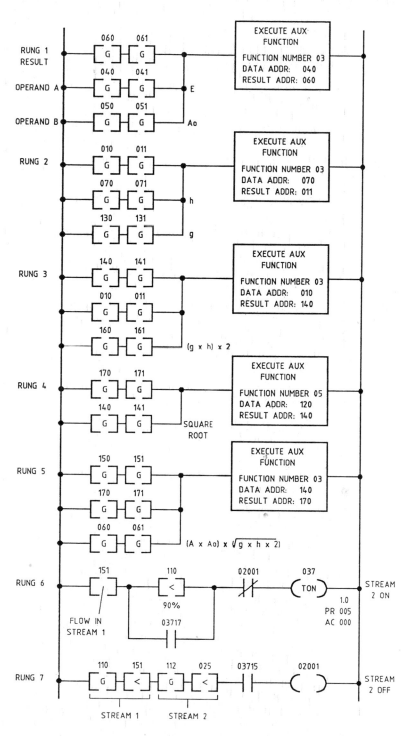

Figure 6.17   Ladder diagram for flow calculation and stream switching

The second section of this application is stream switching. Consider stream 1 is on and has reached approximately 90 per cent of its ability to supply the necessary flow. At this point, the stream 2 actuator will open to share the flow, therefore causing the flow through stream 1 to reduce. The reduction of flow through stream 1 must not allow stream 2 to shut down. If it did, it would cause all the flow to go through stream 1 again which would cause stream 2 to open again so that the valve on stream 2 would continually oscillate open and closed. It is also not necessary to shut down all streams, except in an emergency, as interruption to the gas supply may have severe consequences.

The requirement of the streams when switched is:

- stream 1 switches stream 2 at 90 per cent of stream 1 capacity;
- stream 2 switches stream 3 at 80 per cent of stream 2 capacity;
- stream 3 switches stream 4 at 70 per cent of stream 3 capacity.

Rung 1 multiplies E (operand A) and Ao (operand B) and stores the result at address 060 using function 03.

Rung 2 multiplies h and g and stores the result at address 120.

Rung 3 multiplies by 2 the result of the g and h multiplication in rung 2 by 2.

Rung 4 takes the square root of $g \times h \times 2$ and stores the result at address 170 using function 05.

Rung 5 multiplies the value of the square root of $g \times h \times 2$ by the value of $E \times Ao$ and stores the result, which is the value of Q at address 150. The formula now solved can be displayed output as an analog or digital value.

Rung 6: When the flow in stream 1 is 90 per cent, stream 2 actuator is operated.

Rung 7: Contact 03715 latches stream 2 on rung 7. Stream 2 will only be closed when stream 1 is less than 90 per cent and stream 2 is less than 10 per cent. Contact 02001 releases stream 1 actuator.

A similar configuration of contacts will be required to stream switch each regulator run.

6.5 Numbering system conversions

The microprocessor carries out its internal operations using binary numbers. As it is necessary for the PLC to communicate with the physical world where inputs and outputs may be decimal, and in particular binary coded decimal (BCD), it is often necessary to convert between BCD, decimal, and binary. Input information in a binary format can be entered directly into the programmable controller memory, and be operated on directly, under the control of the user program.

BCD-to-binary and binary-to-BCD conversions are available in most programmable controllers as individual commands or as an arithmetic function. In the Allen-Bradley PLC 2/30 a BCD-to-binary and binary-to-BCD function is specifically provided to carry out the operation. The Allen-Bradley PLC 2/17, however, has the numbering conversions included with the execute auxiliary function (EAF) among the extended arithmetic functions. An example of each programming technique is included in this brief discussion of number conversions.

BCD-to-binary and binary-to-BCD conversion instructions in the Allen-Bradley PLC 2/30 are shown in Figure 6.18. When input 11100 passes power the conversion takes place where any BCD number from 0 to 4095 will be converted to its 12-bit decimal equivalent. The BCD input requires two consecutive data table words to store two 3-digit BCD whole numbers. Bit 14 is used to indicate an overflow has occurred when the BCD number is greater than 4095. The binary result is stored in the lower 12 bits of a user-selected word.

To program the conversions the CONVERT operation is selected from the keyboard, followed by 0 for BCD-to-binary conversion and 1 for a binary-to-BCD conversion.

The function of the abbreviations shown in the conversion ladder diagram of Figure 6.18(a) and (b) are as follows:

1 BCD – binary coded decimal section of the instruction;
2 Addr – address of two-word BCD or single-word binary number;
3 BCD – result or input BCD number;
4 Binary – binary section of instruction;
5 Addr – address of the stored or input binary word;
6 Data – 2-bit binary number result or input to conversion;
7 the overflow bit (OV) indicates when the BCD has exceeded 4095 or the binary number exceeds twelve 1s.

The Allen-Bradley PLC 2/17 has the number conversions included as part of the execute auxiliary function (EAF) instructions, programmed as shown in Figure 6.19.

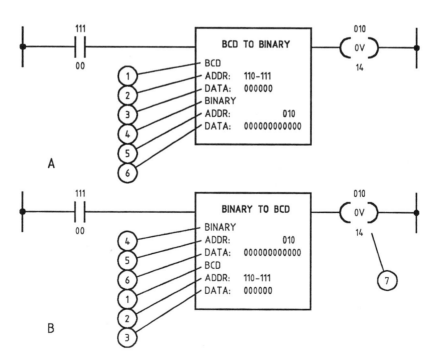

Figure 6.18(a) BCD-to-binary conversion   (b) binary-to-BCD conversion

Rung 1 contains an up counter which is incremented by input 11100. Rung 2 contains the GET function 050 which stores the result of the conversion, and the two-word BCD input 040 and 041. The auxiliary function selected is 13 for a BCD-to-binary conversion and 14 for a binary-to-BCD conversion. Rung 2 containing the EAF is unconditional, i.e. has no preceding conditional input logic and will make the conversion each scan cycle of the processor.

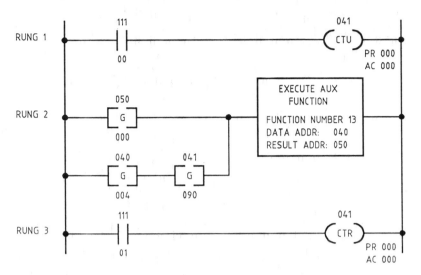

Figure 6.19   Number conversion using the execute auxiliary function

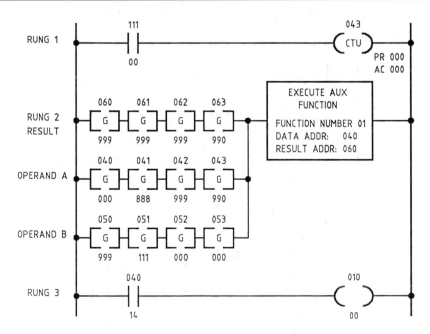

Figure 6.20   12-bit arithmetic using the EAF

Table 6.1   Arithmetic and auxiliary functions available on the Allen-Bradley PLC 2/17.

Function	Number
Addition	01
Subtraction	02
Multiplication	03
Division	04
Square root	05
Average	06
Standard deviation	07
Set clock	10
Set data	11
Set leap year and day of week	12
BCD to binary	13
Binary to BCD	14
Read clock	15
Read data	16
Read leap year and day of week	17
Proportional integral and dirivite (PID)	27
First in first out load	28
First in first out unload	29
Logarithm base 10	30
Logarithm Base e	31
Power of e	32
Powers of v	33
Reciprocal	34
Sine X	35
Cosine X	36
10 raised to power of X	37

6.6 Extended arithmetic functions

The evolution of the programmable controller has led to the development of a wide range of arithmetic functions as well as a number of auxiliary functions. Table 6.1 shows the arithmetic and auxiliary functions available on the Allen-Bradley PLC 2/17.

An example of using the Allen-Bradley EAF is shown in Figure 6.20, where two 12-digit numbers are added. When the PLC is switched to the run mode the data in operand A will be added to the data in operand B. The result of the addition is stored at a separate data table location. If counter 043 is incremented by the operation of input 11100, the data in operand A will be incremented; if an overflow condition is created, bit 14 of 040 in operand A will be set to a logical 1 and output 01100 will become active.

6.7 Test questions

Multiple choice

1 The BCD bit pattern for a decimal number 3429 + 1 is equal to:
   (a) 110101100110.
   (b) 0011010000110000.
   (c) 0011010000101001.
   (d) 1100101111010110.

2 The bit pattern at the output assembly required to cause a seven-segment display to indicate the digit 6 after decoding by a 7448 TTL device is:
   (a) 0011.
   (b) 1100.
   (c) 0110.
   (d) 1001.

3 The execute auxiliary function allows the programmer to:
   (a) execute an internal field function.
   (b) jump to a subroutine.
   (c) produce a one shot timer.
   (d) carry out complex arithmetic.

4 The thumbwheel switch and seven-segment display are:
   (a) digital input and digital output respectively.
   (b) analog input and digital output respectively.
   (c) digital input and analog output respectively.
   (d) analog input and analog output respectively.

5 Which of the following functions can be carried out by using word manipulation?
   (a) Load counter and timers with accumulated values.
   (b) Immediate functions.
   (c) Compare functions.
   (d) Master control functions.
   (e) Arithmetic functions.

True or false

6 A thumbwheel switch is a digital device that requires a word to input the data.

7 Seven-segment displays can be driven directly from a PLC digital output assembly.

8 The Allen-Bradley GET BYTE and LIMIT TEST compares a decimal value to a lower or upper byte of a 16-bit word.

9 Arithmetic instructions are word instructions.

10 Overflow bits are used when carrying out number conversions to indicate if the result exceeds a predetermined number.

Written response

11 Design a program that will compare two BCD numbers and make an output active if number A is greater than or equal to number B.

12 Design a program that will convert a BCD number into its binary equivalent.

13 Design a PLC program that will multiply two numbers, divide the result of the multiplication by an input from a thumbwheel switch, and compare the result with two set points outside which a high and low alarm will result.

14 State the advantage obtained by a programmable controller being able to handle functions in the form of data words.

15 Design a PLC program and describe the hardware which would allow a PLC to accept an input from a thumbwheel switch and display the numerical value of the thumbwheel switch on a seven-segment display.

Process control and data acquisition

7.1 Introduction to process control and data acquisition

In modern manufacturing industries there are four types of processes being carried out which can be categorized as:

- continuous process;
- batch production;
- individual products production;
- data acquisition.

Each of the four processes may be carried out by using a programmable controller as the controlling element in the operation.

The characteristic of the continuous process is that raw materials are fed into a continuously running system and finished product is the output. The raw materials enter the system as a steady stream, are usually processed in the same way, and often the run of materials cannot individually be tracked through the system. An example of a continuous process is a plastic blow-moulding plant. Once the blow-moulding process commences it is continuous for a relatively long period of time, typically several hours to several days. The raw materials are continuously fed into the system via an automatic feeding system, and output finished products are stored. Figure 7.1(a) shows a block diagram of a continuous process system.

The batch process has set, identifiable, discrete loads of raw materials which are sequentially introduced to the process. The process is carried out, and the finished

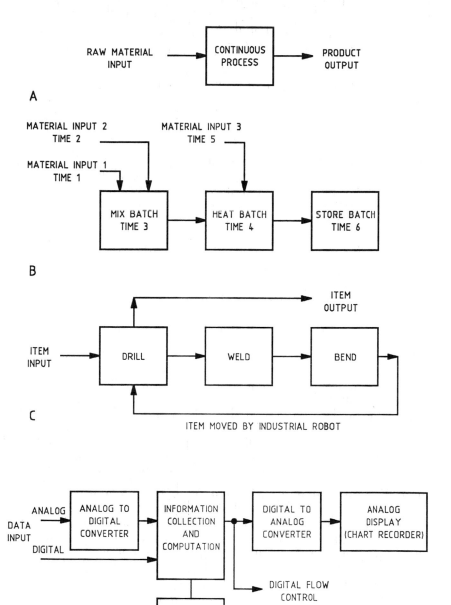

Figure 7.1(a) continuous process (b) batch process (c) individual (d) data acquisition system

product, which may undergo further processing, is stored and another batch of product is produced. Each batch of product may be different. An example of a batch process is a glue-making process where several chemicals are added together, mixed, heated, more chemicals are added, processed, and then stored. Each batch of glue made may have differing characteristics by design. Many chemically-based products are manufactured by using batch processes. Figure 7.1(b) shows the block diagram of a batch process system.

The individual product production process causes a single item of product to proceed through a number of operations prior to being output as a completed item. The item being produced may be required to be bent, drilled, welded, etc. at the steps in the process. An industrial robot, in conjunction with a computer numerical control (CNC) machine, can be used in an individual item process. Figure 7.1(c) is a block diagram of an individual product production process where an item is drilled, welded, bent, then drilled again.

The acquisition of data is included as a type of process because it is used in industry to measure, check, define, or reject a product. In many applications it is the sole function of large sections, if not the entire plant. All processes have some elements of data acquisition included within them, but it would be reasonable to expect a petrochemical metering station, with the purpose of measurement and flow control, to accurately record and measure the input data, and from the input information control the flow of the station output. Figure 7.1(d) is the block diagram of a data acquisition system.

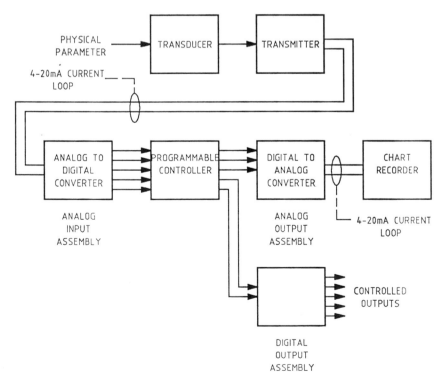

Figure 7.2   Data acquisition system

Data acquisition is a relatively new concept applied to programmable controllers. As the new generation of programmable controllers have been designed to accept a wide range of inputs, both analog and digital, and have powerful arithmetic functions and controlling ability, they are now suitable for some data acquisition applications.

The data acquisition system consists of a number of field transducers connected to transmitters; a digital-to-analog converter interfaces to the programmable controller. The programmable controller uses the analog input information to control the process and provide an analog output to a permanent storage device such as a chart recorder. A block diagram of a data acquisition system using a programmable controller is shown in Figure 7.2. The method used to connect analog inputs and outputs is the same for all processes.

7.2 Transducers

A transducer is an energy conversion device which converts input variables from the physical world into electrical signals suitable for measurement and control. Transducers are divided into two types: primary, and secondary. Primary devices are for the measurement of:

- linear movement;
- angular movement;
- temperature;
- illumination;
- time;
- force.

Many secondary devices can be developed from the primary device to measure quantities derived from the primary measurement which utilize intermediate stages. A number of transducers have been selected for discussion to convey the principles of operation of this diverse range of devices. The passive transducers in direct contact with the process that will be discussed are:

- light-dependent resistors;
- thermistors;
- capacitance proximity detectors;
- strain gauges.

The secondary transducers discussed will be:

- linear variable differential transformers;
- shaft position and speed encoders;
- optical and infra-red detectors;
- pressure measuring devices.

Light-dependent resistors (LDR)

Light-dependent resistors are non linear devices which will change resistance depending on light intensity. Typically, the dark resistance of an LDR is in excess of 1 million

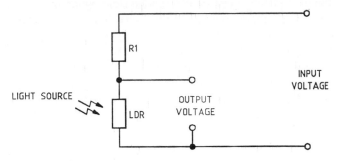

Figure 7.3   Light dependent resistor

ohms, and in bright light the resistance drops to 150 ohms; this means the LDR is sensitive to small changes in light intensity. The disadvantage of the LDR as a transducer is its non linearity and slow recovery rate making it unsuitable for applications where rapid changes in light intensity are expected. Figure 7.3 is a light-dependent resistor in a circuit where changes in light intensity can be used to produce an output voltage.

Thermistors

The thermistor is a non linear device where conductivity varies with changes in temperature. Thermistors either have a negative temperature coefficient of resistance where an increase in temperature will cause a decrease in resistance, or a positive temperature coefficient of resistance where an increase in temperature will cause an increase in resistance. Figure 7.4 shows a circuit where a negative temperature coefficient thermistor is used to sense temperature changes and produce an output voltage. As the temperature of the thermistor changes, the resistance change causes the voltage at the inverting input of the operational amplifier to change, producing a corresponding change in output voltage.

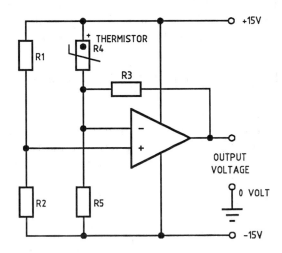

Figure 7.4   Themistor

Capacitance proximity detector

A capacitance proximity detector can be used as a simple, effective, and accurate means of measuring the closeness of objects or levels of substances. As capacitance is proportional to the dielectric material and inversely proportional to the distance between the capacitor plates, capacitance changes can be used to produce an output. Level can be measured by capacitance changes as a substance replaces the dielectric. Proximity can be detected as the capacitance plate distance is varied by one capacitor plate being in close or distant proximity to the other. The capacitive transducer can detect changes in capacitance of a few pico farad.

$$C \propto \frac{Ak}{d}$$

where
C = capacitance in farads
A = area of the plates
d = distance between the plates
k = dielectric constant

Strain gauges

The strain gauge is a type of transducer used to measure force. The principle of operation of the strain gauge is based on a conductor's resistance changing as the dimensions of the conductor change. The change is represented in the formula:

$$R \propto \frac{l\rho}{a}$$

where
R = resistance in ohms
l = conductor length
a = cross-sectional area
ρ = specific resistance of the material

If a force is applied to a strain gauge, causing the conductor's length and therefore cross-sectional area to change, the resultant change in resistance can be detected. The conductor is resistance wire, zigzagged into a grid formation to obtain the length required to provide a measurable resistance change. A two-dimensional strain gauge and circuit suitable to provide a voltage output from the strain gauge transducer, are shown in Figure 7.5.

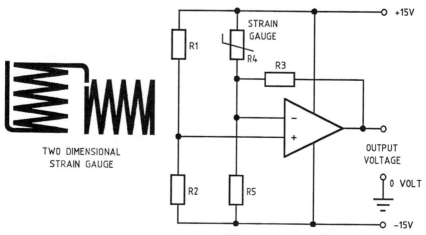

Figure 7.5   Strain gauge

Linear variable differential transformers (LVDT)

The LVDT consists of a center-tapped solenoid which has a moveable hard steel core. The solenoid can be energized directly or by a separate primary coil. As the core moves with respect to the solenoid winding, the flux lines cutting the winding will vary, resulting in an imbalance between the two halves of the center-tapped solenoid, the amount of unbalance being linearly related to the core movement. Linear variable differential transformers are used in industry in pressure transducers, load cells, and strain indicators. Figure 7.6 shows the construction of a LVDT which is electrically excited from an external AC source connected to the primary solenoid and has two secondary windings between which the difference voltage is generated.

When the core of the LVDT is exactly at the center of the solenoid, the output is zero. Any shift in the core results in an output voltage being generated.

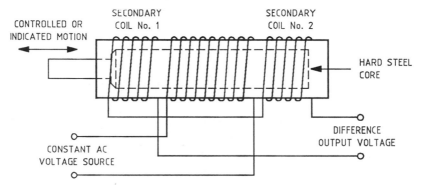

Figure 7.6   Construction of LVDT

Shaft position and speed encoder

There are a number of techniques used to determine the position of a shaft, or the speed at which a shaft rotates. A simple means is to employ a light source and light-sensitive diode in conjunction with a slotted disk connected to the motor shaft. As the slot moves between the light source and the light-sensitive diode, the resistance of the diode will reduce. The change of diode resistance can be fed into an electronic circuit which counts the revolutions per second for the speed indication. Alternatively, the shaft would be in a known position each time the light-sensitive diode's resistance reduced.

In many applications, the slot in the disk can be replaced by a reflective spot on the surface of the disk, thus allowing the light source and detector to be located on the same side of the disk. Similar results can be obtained by placing a reflective spot or code directly on the rotating shaft. Figure 7.7 shows a slotted disk used for a revolutions-per-minute (RPM) counter where the frequency of the output pulses will vary, depending on the RPM of the shaft.

The slots or reflective spots are often used in the shaft encoder in such a way as to form a 4-bit code. The code used is termed a Gray code. Gray codes are designed in such a way that only one bit changes from one step to the next, thus reducing the possibility of incorrect results. An example of the Gray code is shown in Table 7.1 where it is compared to a binary code. In the binary code, when the transition between 3 (0011) and 4 (0100) takes place, the number read by an external device could be 6

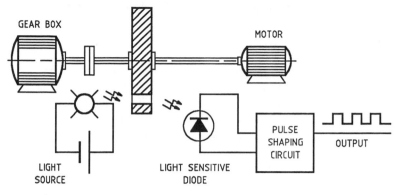

Figure 7.7 Rev counter

(0110) if the second bit of the 0011 is late in changing. Using the Gray code, only one bit changes in any transition, therefore the number read can only be either 3 (0010) or 4 (0110).

Table 7.1 Comparison of Gray code with binary code

Decimal	Binary	Grey
0	0000	0000
1	0001	0001
2	0010	0011
3	0011	0010
4	0100	0110
5	0101	0111
6	0110	0101
7	0111	0100
8	1000	1100
9	1001	1101

Optical and infra-red detectors

Optical and infra-red detectors have many applications as transducers. The previously discussed shaft encoder employed an optical technique to obtain the position data using a light-sensitive diode which is an optoelectronic device. Other devices which fit the optoelectronic category are:

- photo or light sensitive transistors;
- optical couplers;
- light-emitting or light-sensitive diodes, producing both visible and infra-red light;
- laser and laser diodes;
- fiber optics;
- photovoltaic cells (solar cells).

Optoelectronic devices are used in many applications such as level measurement, density measurement, level and alignment measurements, and alarm systems.

Infra-red devices are used to detect warm or hot objects including plant operating personnel. They are also often used as safety beams to define machine boundaries and

135

stop a machine if personnel or any other detectable objects are inside the machine perimeter.

Pressure measuring devices

Process industries often require pressure to be measured and there are many transducers which can be used for this. Some pressure measurement devices have previously been discussed, for example the strain gauge. Elastic elements such as displacement bellows and Bourdon tubes are simple transducers which are often used in industry as pressure transducers. Figure 7.8 shows a Bourdon tube connected to a potentiometer to produce a voltage output.

Piezoelectric transducers and load cells are also used as pressure transducers, as well as manometers, in conjunction with optical sensing devices, to detect the level of a column of water or mercury.

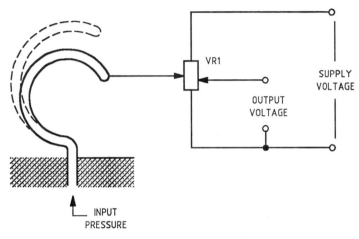

Figure 7.8   Bourdon tube

7.3 Transmitters

The transducer signal is often not suitable to transmit information from the field to the programmable controller. For example, the signal may be non linear and therefore not suitable for analog-to-digital conversion, or only a few millivolts and perhaps susceptible to electrical interference. A transmitter is therefore used in conjunction with the transducer to convert the transducer signal to one which can transmit information to the input assembly of the programmable controller. The transducer and transmitter combination must produce an output which has required characteristics, specified by the system designer, under the following headings:

- repeatability – obtaining a similar resultant output each time the instrument is tested; the scatter around the mean of results can be expressed as standard deviation;
- resolution – the smallest input quantity change that the instrument is able to detect;
- precision – how well the instrument measures and outputs a reliable signal;
- accuracy – the ability of the instrument to be correct.

The output of the transmitter will be a voltage or current which will represent the range of input signal being measured. The typical voltage and current transmitter outputs are:

- 4–20 mA;
- 10–50 mA;
- 0–5 volt;

- 1–5 volt;
- 1–10 volt.

To obtain the required output voltage or current the transmitter will contain an amplifier as well as a filter to assist in reducing unwanted electrical noise.

An example of a pressure transmitter is one that may measure 0 to 100 psi which will represent an output of 4 to 20 mA as shown in Table 7.2, therefore producing an output current for any value of input pressure over the range of 0 to 100 psi. This type of output signal is termed an analog signal.

Table 7.2　Input pressure/output current of a pressure transmitter

Input pressure (psi)	Output current (mA)
0	4
25	8
50	12
75	16
100	20

A transmitter may be ranged to suit a particular application. For example, if a temperature of between 95°F to 105°F is critical for the process involved, the transmitter and analog input assemblies can be ranged accordingly, as shown in Table 7.3.

The type of transmitter output, voltage or current, is dependent on the application. The voltage output transmitter is satisfactory if the input to the programmable controller analog assembly is within a few yards or meters of it, as long cable runs will introduce a voltage drop along the cable which must be compensated for at the analog input assembly. If the transmitter is located some distance from the programmable controller analog input assembly, a current loop is preferred, where 4 to 20 mA is used to express the transducer.

Table 7.3　Input temperature/output current of a temperature transmitter

Input temperature (F)	Output current (mA)
95	4
97.5	8
100	12
102.5	16
105	20

The current loop is less susceptible to electrostatic or electromagnetic induction than the voltage loop. The current loop also has the advantage that it can be used to detect a loop which may have an open circuit fault condition by using 4 mA to represent 0, or the lowest measured value. A zero volt for lowest measured value would be within range if the circuit were open and therefore would not be detected. The zero volt output transmitter is likely to suffer from induced noise at the low end of its range due to electrostatic and electromagnetic induction produced in the industrial environment.

Figure 7.9 shows an industrial process where 4 to 20 mA signals are used to represent pressure, temperature and flow. The blocks which go to make up the analog input assembly include an analog multiplexer, sample and hold, and an analog-to-digital converter which, together with the transducer, transmitter and a microprocessor-based device, make up a data acquisition system. As the analog input assembly accepts a voltage signal, the 4 to 20 mA current loop is fed into an input precision resistor to convert the current to a suitable voltage. The 4 to 20 mA signal can be converted to 1 to 5 volt by using a 250-ohm resistor. The resistor would have a tolerance of 0.5 per cent or better to make sure that the voltage across the resistor, which is caused by the current through it, is consistent.

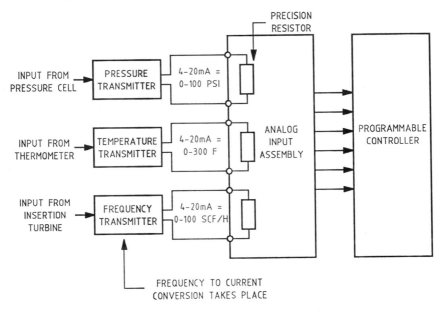

Figure 7.9   Analog inputs to the programmable controller

Data acquisition systems

Figure 7.10 shows the block diagram of a data acquisition system. The transducer and transmitter have been discussed above. The analog multiplexer, sample-and-hold circuit, analog-to-digital converter and the digital-to-analog converter will be covered here.

The analog multiplexer is an electronic device which presents a voltage signal to the sample-and-hold circuit. The sample-and-hold circuit has an output which is in direct relationship to the current or voltage provided by the transmitter. The multiplexer is a time division type which switches each analog loop to the sample-and-hold circuit for a predetermined length of time.

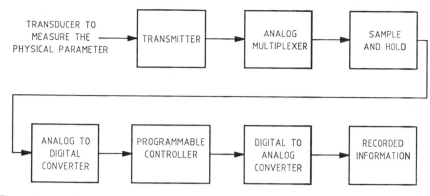

Figure 7.10   Block diagram of a data acquisition system

The sample-and-hold circuit acquires the voltage of the output of the analog multiplexer and retains this voltage in its output for the time it takes for the analog-to-digital converter to convert the signal to its digital equivalent. The converter places the digital word into the processor memory.

The timing and sequencing of the events in the data acquisition system are carried out by the PLC processor. Under the control of the user program, the programmable controller can present a word of digital information to a digital-to-analog converter, which will convert the digital signal to its analog equivalent. The resultant analog equivalent can then be used to control an output device such as an actuator, or it can be recorded.

7.4 Data acquisition theory

Three areas of data theory encompass the fundamentals of data acquisition:

- quantizing theory;
- sampling theory;
- converter codes.

To convert the analog signal to its digital equivalent, a two-step process is required. The first step is to convert the analog signal to a discrete output value; this is termed quantizing. The second step is to change the discrete output value to a usable code of a digital format.

Figure 7.11 shows the development of quantizing theory as applied to a 4-bit ADC with an analog range of 0 to 5 volts. The ADC uses sixteen steps to convert the 1 to 5 volt signal to a 4-bit code.

The ADC has a resolution of the number of output values expressed in bits; thus the 4-bit ADC has a resolution of 4 bits which is equal to 16. The quantizing process has an inherent quantizing error the size of which is directly related to the number of discrete values. Consider the values in Table 7.4 which are derived from the 4-bit ADC discussed in Figure 7.11. Each output code has an equivalent center point input voltage. A defined voltage either side of the center input voltage will be recognized as the same code as the center point input voltage and therefore will constitute the quantizing

139

error. In this case the quantizing error is equal to $+Q/2, -Q/2$, where Q = quantization size which can be calculated.

$$Q = \frac{\text{Full scale resolution}}{2^n}$$

$$= \frac{5}{16}$$

$$= 0.3125$$

Table 7.4   4-bit ADC conversion voltages

Output value	Output code	Center point I/P voltage	Quantizing error	
			Lower limit voltage	Upper limit voltage
16	1111	4.6875	4.375	4.84375
15	1110	4.375	4.0625	4.6875
14	1101	4.0625	3.75	4.375
13	1100	3.75	3.4375	4.0625
12	1011	3.4375	3.125	3.75
11	1010	3.125	2.8125	3.4375
10	1001	2.8125	2.5	3.125
9	1000	2.5	2.1875	2.8125
8	0111	2.1875	1.875	2.5
7	0110	1.875	1.5625	2.1875
6	0101	1.5625	1.25	1.875
5	0100	1.25	0.9375	1.5625
4	0011	0.9375	0.625	1.25
3	0010	0.625	0.3125	0.9375
2	0001	0.3125	0.15625	0.625
1	0000	0	0	0.15625

Here, 2^n is the number of converted bits, and in the case of the 4-bit ADC under discussion, each bit is equal to 0.3125 volts. The quantization error can be reduced by increasing the number of conversion bits, therefore increasing the resolution by making the resolution bits smaller.

Sampling theory

An analog-to-digital converter requires a finite time to perform the quantizing function and then convert the quantized value to a usable code. During the time it takes to quantize and code, the input signal may change value with respect to time; therefore the accuracy of the converted signal is dependent on conversion speed. Figure 7.12 shows how the input signal may change during sampling. The time taken to convert the signal is termed the aperture time, or aperture window, and can be expressed as the formula shown. The sampling error will be greatest when the input signal is changing at the fastest rate with respect to time, therefore sample 2 will have a greater amplitude uncertainty than sample 1.

$$\triangle V = ta\ \frac{dv(t)}{dt}$$

where $\triangle V$ = sampled signal amplitude uncertainty

ta = aperture time

dv/dt = rate of change with respect to time of the sampled input signal

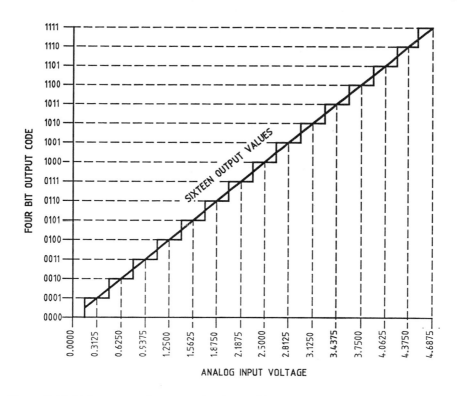

Figure 7.11   Voltage to code conversion for a 4 bit analog to digital converter (ADC)

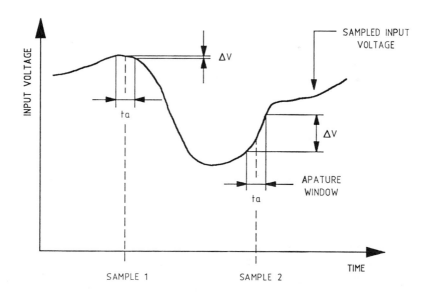

Figure 7.12   Sampling conversion amplitude uncertainty

Converter codes

As the microprocessor-based device is a digital device, digital codes must be used. The digital system is considered to have a number of advantages over an analog system:

- A digital system is less susceptible to noise. It has two states, 1 and 0, whereas the analog system has a continuously varying state which cannot distinguish between unwanted noise signals and the required signal.
- A digital system is less complex to adjust. The analog system requires frequent and accurate calibration and is susceptible to output changes with temperature; the digital signal, when set up correctly, requires minimal calibration. Digital systems are also less complex to design.
- Mathematical functions can be carried out more efficiently in the digital system.
- The digital system is considered to be more precise than the analog system over long-term operating periods. This is because the digital signal is reliably correct to within one bit of the device's resolution.

The most common code used in conversion is the binary code which is a positive-weighted code containing no negative conversions. A binary code in a converter represents a fraction of the full scale value of the converter. An 8-bit binary code has 2^8 or 256 states. The percentage resolution of the 8-bit code equals:

$$\% \text{ resolution} = \frac{1}{2^n - 1} \times \frac{100}{1}$$

$$= \frac{1}{2^8 - 1} \times \frac{100}{1}$$

$$= 0.39\%$$

An 8-bit binary code of 10100111 would represent a decimal fraction of a 0.65234375 calculated as shown in Example 1. Table 7.5 shows weighting of the 8-bit binary fraction shown in Example 1. The least significant bit has an analog equivalent of Q which can be calculated:

$$Q = \frac{\text{Full scale resolution}}{2^n}$$

```
1 × 0.5          = 0.5
0 × 0.25         = 0.0
1 × 0.125        = 0.125
0 × 0.0625       = 0.0
0 × 0.03125      = 0.0
1 × 0.015625     = 0.015625
1 × 0.0078125    = 0.0078125
1 × 0.00390625 = 0.00390625
                   0.65234375
```

Example 1

The value of the digital code represents full scale $(1 - 2^{-n})$, therefore, in the 8-bit converter, with an output range of 0 to 10 volts, the maximum code is 11111111 and a maximum analog value of $10(1 - 2^{-8}) = 9.9609375$ volts. The converted analog never reaches the defined full scale value.

Table 7.5   Weighting of 8-bit binary fraction

LSB				MSB			
Binary code	1	0	1	0	0	1	1	1
Binary weighting	2^{-1}	2^{-2}	2^{-3}	2^{-4}	2^{-5}	2^{-6}	2^{-7}	2^{-8}
Fraction	$\frac{1}{2}$	$\frac{1}{4}$	$\frac{1}{8}$	$\frac{1}{16}$	$\frac{1}{32}$	$\frac{1}{64}$	$\frac{1}{128}$	$\frac{1}{256}$
Decimal fraction	0.5	0.25	0.125	0.625	0.3125	0.015625	0.0078125	0.00390625

7.5 Analog-to-digital conversion

There are many circuit configurations that will convert an analog signal to its digital equivalent; two of the most common conversion methods will be discussed in this section: dual slope integrator, and successive approximation.

Dual slope integrator

Figure 7.13 is a block diagram of a dual slope integrator analog-to-digital (A/D) converter, which consists of six separate sections: voltage reference, integrator, comparator, control logic, counter, and clock. The conversion commences when a start conversion signal resets the counter to zero and discharges the integrator capacitor C. The analog input voltage is then presented to the input via the switch, causing an input current to flow, thus charging the integrator capacitor and causing the output voltage of the integrator to rise. When the analog input is presented to the input, the comparator switches, producing a positive output which enables the counter. The counter is allowed to count until it overflows. The counter overflow signal switches the switch from the analog input to the negative reference voltage, discharging the integrator capacitor. The clock pulses are counted until the comparator detects the zero crossing, at which time the counter is stopped, and an end-of-conversion signal is generated. This signal indicates that the output of the binary counter is the converted digital signal.

Figure 7.14 shows the timing sequence of the dual slope integrator A/D converter. Time $t1$ is the fixed time taken to cause the counter to overflow. Time $t2$ is the measured time taken to discharge the integrator capacitor, and is proportional to the analog input voltage. Two analog input signals are being converted. As the second conversion presents a larger voltage to the integrator input, it takes longer to discharge the integrator capacitor and therefore the counter counts to a higher value than in conversion 1. As the digital output represents the ratio of the input voltage to the reference voltage in time, the measured time can be calculated:

$$t2 = t1 \frac{\text{Input voltage}}{\text{Reference voltage}}$$

where $t1$ = fixed time
$t2$ = measured time

143

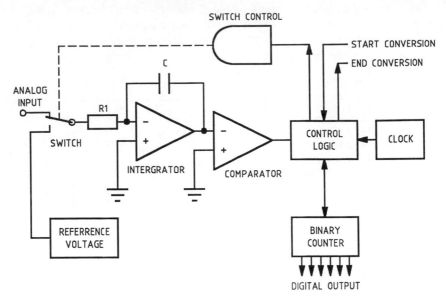

Figure 7.13   Dual slope integrator analog to digital converter

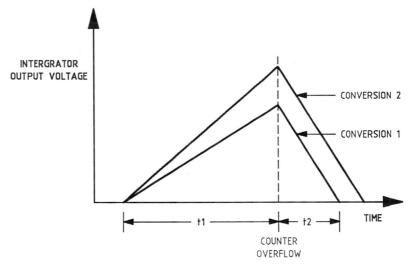

Figure 7.14   Integrator waveform for the dual slope A/D converter

Successive approximation

The successive approximation method of conversion uses a digital-to-analog converter in the feedback loop. The operation of the successive approximation A/D converter
is based on the device making several attempts to compare the analog input voltage
with an analog reference voltage generated by the digital-to-analog converter. The advantage of the successive approximation converter is that the conversion is completed
in $n + 1$ clock cycles or less, n being the resolution of the converter. This makes this

converter relatively fast; it is therefore used in applications where conversion speed is important.

Figure 7.15 is the block diagram of the successive approximation A/D converter. The operation of the converter is such that the successive approximation register controls the converter by implementing an output voltage equal to the most significant bit. This voltage is fed into the comparator where it is compared to the input analog voltage. The most significant bit is equal to half full scale voltage of the digital-to-analog converter. If V input is greater than V output of the digital-to-analog converter, the next most significant bit is set and the analog output of the converter is increased from half to three-quarters of full value. A comparison takes place, and, if the compared values are less than three-quarters of full scale, the comparator turns off the second most significant bit, turns on the third most significant bit, and another comparison is made. After a number of comparisons, the digital outputs of the successive approximation register, which are latched to a logical 1, are the digital equivalent of the analog input. The output action of the successive approximation A/D is shown graphically in Figure 7.16.

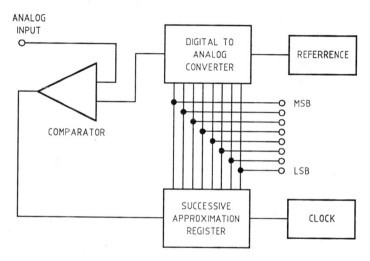

Figure 7.15 Successive approximation A/D convertor an 8 bit successive approximation A/D convertor

7.6 Digital-to-analog conversion

A digital-to-analog converter A/D produces an output voltage or current with an amplitude which is directly related to the value of the digital input. The digital signal format presented to the input of the D/A is usually parallel, where all the input bits are handled together. Of the many types of D/A, the weighted current or voltage type is popular.

The R–2R D/A converter is shown in Figure 7.17. Two resistors, one twice the value of the other, are used in a ladder configuration. The operation of the R–2R ladder network is based on the binary division of current flowing in the ladder network made

145

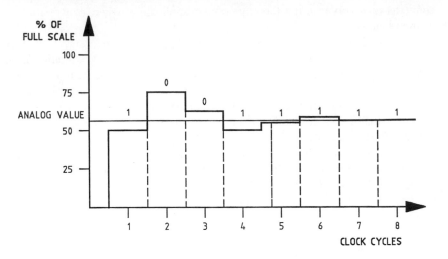

Figure 7.16   Successive approximation A/D convertor conversion steps

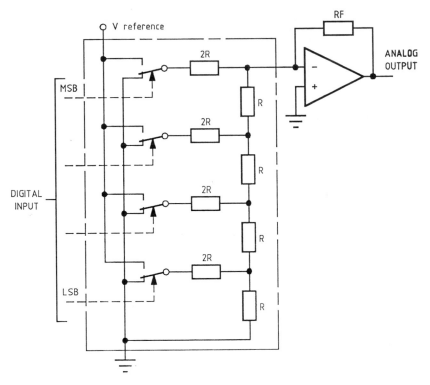

Figure 7.17   R–2R ladder D/A converter

up of series resistors with a value of R and shunt resistors with a value of 2R. The current produced by the MSB, being a logical 1, is twice the current of the second significant bit, being a logical 1. The current produced by the second significant bit is twice

146

that of the third, and so on. An electronically-controlled switch switches the input current between the summing lines and ground. The 2^{-n} represents the input currents flowing through the 2R resistors. The output current of the summing lines is the input to the operational amplifier which converts the input current to a voltage that is the analog equivalent of the digital input. If all the digital bits are 1, the ladder output current is equal to:

$$I \text{ ladder out} = \frac{V \text{ reference}}{R} (1 - 2^{-n})$$

7.7 Operational amplifiers

The operational amplifier is found throughout the process industry. The circuit properties that can be developed by using an operational amplifier (op-amp) are:

- wide bandwidth;
- high gain;
- high input impedance;
- low output impedance.

The operational amplifier can operate as an unbalanced or differential input amplifier and be used to amplify input signals which can be either AC or DC depending on the PLC input signal requirements. Figure 7.18 shows an unbalanced inverting DC operational amplifier operating from a split rail power supply. The operational amplifier has positive and negative inputs, therefore allowing an inverting or non inverting function at its input. In Figure 7.18, the non inverting input, indicated by a plus sign (+), is connected to ground, and the signal is connected to the inverting input, indicated by a minus sign (−). If one side of the input signal is connected to earth in an amplifier configuration it is termed an unbalanced input. The gain of the amplifier is determined by the value of feedback resistor RF in relation to the input resistor Rin. The value of the voltage gain of an operational amplifier can be expressed mathematically:

$$AV = \frac{RF}{Rin}$$ where AV = voltage gain
RF = feedback resistance in ohms
Rin = input resistance in ohms

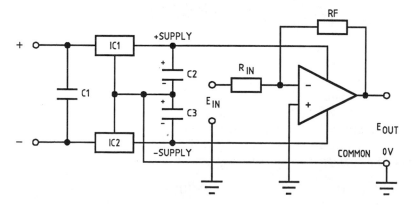

Figure 7.18   Inverting operational amplifier with split rail power supply

The output voltage of the amplifier can be expressed as:

$$E\text{out} = -AV\,E\text{in}$$

where $E\text{out}$ = output voltage

$-AV$ = voltage gain of an inverting amplifier

$E\text{in}$ = input voltage

The split rail power supply is obtained by using a positive, three-terminal regulator integrated circuit (IC1) and a negative output three-terminal regulator (IC2) connected in such a way as to produce a positive voltage output and a negative voltage output with respect to a common rail.

An offset control is often required when the analog input to the PLC analog module is 4 to 20 mA. The offset control allows the base line of the operational amplifier to be shifted to produce zero output voltage when an input current of 4 mA is flowing. The gain of the amplifier is determined by the feedback resistor (RF), and input resistor. The gain is maximum when RF is maximum, and minimum when RF is minimum. The value of RF is determined by the voltage required at the output of the operational amplifier. Figure 7.19 shows an unbalanced input operational amplifier with offset and gain control.

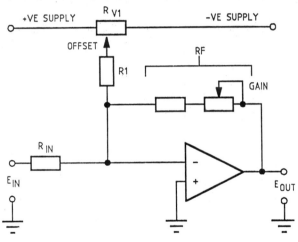

Figure 7.19 Gain and offset adjustment

Differential amplifiers

The differential input amplifier has the input presented to the positive and negative input terminals simultaneously. The advantage of the differential over the unbalanced operational amplifier is that it offers the input control loop common-mode rejection. The differential amplifier with a high common-mode rejection ratio (CMRR) is one which will respond to the difference voltage at its input terminals and not to any voltage which is common to the terminals. Common-mode rejection ratio is a value, expressed in decibels (dB), of the ratio of the differential voltage gain to the common-mode voltage gain. The common-mode rejection ratio can be expressed in the formula:

$$\text{CMRR} = 20 \log 10 \; \frac{AVD}{AVC}$$

where CMRR = common mode rejection ratio in dB

AVD = voltage gain differential

AVC = voltage gain common mode

148

Figure 7.20 is the circuit diagram of a single differential amplifier. If no input signal is present and the field wiring has a voltage induced into it due to a noisy environment, the amplifier will reject the noise voltage at its input. Difference voltages produced by the input signal will be amplified.

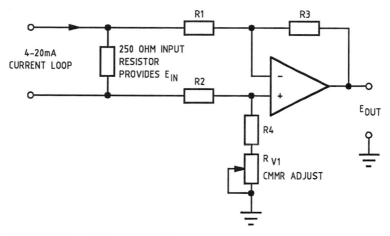

Figure 7.20   Differential input amplifier

Differential instrumentation amplifier

Figure 7.21 shows the schematic circuit diagram of an instrumentation amplifier which will provide the following characteristics:

- high input impedance of 300M ohms;
- high CMRR of at least 100dB;
- gain of between 1 and 1000.

The characteristics of the instrumentation amplifier make it ideal for scientific, industrial and process applications. Differential instrumentation amplifiers are often used as the input amplifiers for programmable controller analog input assemblies.

7.8 System control

Control, as applied to a machine or process, indicates that the result produced is known for a given set of operating parameters. The control may be provided electronically, hydraulically, pneumatically, and so on. The control discussed in this chapter will be electronic, using a programmable controller.

A controlled machine or process implies that the machine or process is being operated by a user-designed system and the output of the machine or process is supervised to detect errors and/or defects in the product. In the case of a programmable controller, the process or machine is operated and supervised under the control of the user program.

The controlled process or machine will contain:

- set points, which are the limits within which the process or machine must operate;

149

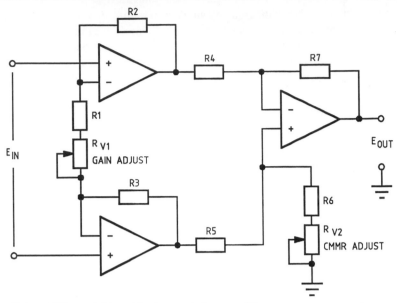

Figure 7.21   Differential input instrumentation amplifier

- process variables, which are continually monitored varying inputs and will determine the required steps necessary to produce the desired output;
- process errors, which are uncontrolled, resulting in the output being modified each time the process or machine carries out its function. The modified output must remain within a specified tolerance. Out of tolerance errors will cause the process or machine to stop producing an output.

Processes can be categorized into two types: open loop, and closed loop. Open-loop processes are not provided with feedback from the process result, and therefore assume, at the conclusion of the user program, that the desired result has been achieved from a known set of input conditions. Open-loop control reduces system complexity and costs less when compared to closed-loop control. Open-loop control, however, has an element of uncertainty and is not considered to be as accurate as closed-loop control.

Closed-loop control contains feedback, that is, some information derived from the process result or output is fed back to the input to indicate that the input signals are having a result on the output. Closed-loop systems are considered to be complex, expensive, and more difficult to adjust or calibrate than open-loop systems, but they have a higher certainty that the process output is occurring as well as an indication of the output accuracy.

An example of an open-loop control system is a stepping motor being driven from a programmable controller. The stepping motor is a permanent magnet motor with a number of sets of coils, termed phases, located around the rotor. The phases are wired to the PLC output assembly and are energized in turn under the control of the user program. The rotor turn is determined by which phases are turned on. The angle of the turn is termed the step angle. The programmable controller does not receive feedback from the motor to indicate that rotation has occurred, but it is assumed that the motor

150

has responded correctly because the motor phases are driven in a sequence. Some information on the motor speed and shaft position can be obtained by using the programmable controller to count pulses when the motor phases are energized.

An example of a motor in a closed-loop system is one where a speed and position analog or digital signal is generated and fed back to the programmable controller where, as the motor is driven, the difference between the actual and required position is detected as an error, therefore causing the motor to drive until the error is zero. The closed-loop system has the following disadvantages:

- There is some transport delay – the time taken for the controlling device to calculate the new position and output the required motor drive signal.
- As analog-to-digital and digital-to-analog conversion are required, quantization errors occur in the system.

These disadvantages are minor and are outweighed by the advantages of process output certainty and accuracy.

The devices used to produce motor feedback are tachometers, both analog and digital, and encoders. The analog tachometer is a DC generator connected to the motor that produces an output analog voltage which is dependent on the speed of rotation of the motor shaft. The derived analog signal is fed into the programmable controller via the analog input assembly.

In its simplest form, the digital or pulse tachometer consists of a slotted disk mounted on the motor shaft. As the disk rotates, light is passed through the slots. Light-sensitive devices are used to detect the light and produce an electronic pulse train with a frequency dependent on the speed of rotation of the disk.

Shaft encoders with a Gray code output are often used for motor control feedback, particularly with robots, as their output allows absolute angular and distance measurement to be made. The shaft encoder can be similar in construction to the pulse tachometer. The Gray code is a code which is designed so that one bit only is changed from one step to the next, thus eliminating the possibility of the type of error that may occur when a code such as BCD is used.

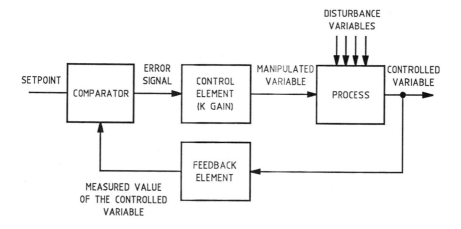

Figure 7.22   Closed loop control

Controller responses

There are four types of control response used in the process industry:

- on/off;
- proportional (P);
- integral (I);
- derivative (D).

Consider the block diagram of the closed loop control system shown in Figure 7.22. A setpoint is compared with a fed back signal, termed the controlled variable, and the difference between the setpoint and the controlled variable signal, termed the error, is used as a signal to the control element to change the process signal. When the error signal is zero, the setpoint has been reached. The disturbance inputs can cause the control variable to change and are unwanted inputs.

On/off control

On/off control, sometimes called two-position control, has two outputs which are full on for maximum output, or full off for minimum output. Figure 7.23 shows a system using on/off control, where a liquid is heated by using steam. The temperature of the liquid to be heated must be above a minimum and below a maximum, therefore upper and lower setpoints are required. When the system is operated, a steam valve opens and applies maximum steam, heating the liquid until the upper setpoint is reached. At this point the steam is shut off. The steam valve will remain closed until the liquid's lower set temperature point is reached, at which time the steam valve will fully open, heating the liquid. The on/off cycle will continue for the duration the system is operating.

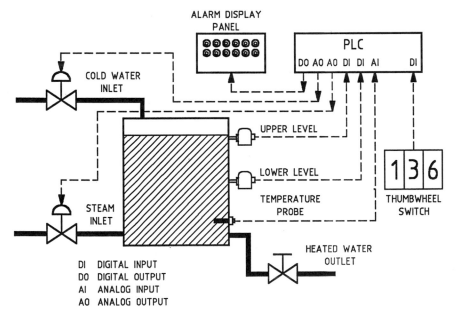

Figure 7.23   Liquid heating system

The programmable controller used in the application shown in Figure 7.23 provides the open and closed signal for the steam valve actuator and the cold water inlet. The steam valve actuator is controlled by the setpoint generated by the thumbwheel switch. Compare functions are used to generate the upper and lower setpoint signals for the PLC. The cold water inlet valve actuator is controlled by the PLC and the water level kept by upper and lower level limit switches. The PLC is also used to generate alarms.

On/off control has a number of disadvantages. Figure 7.24 shows a graphic presentation of the result of on/off control: the temperature will overshoot and undershoot the setpoints; it cannot be maintained at a constant, and the controller will hunt between the setpoints and not settle.

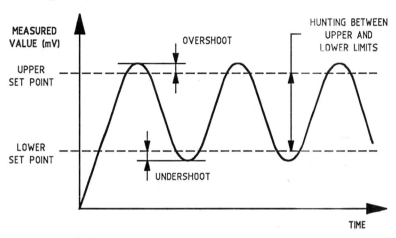

Figure 7.24   On/off control

Proportional control

To overcome some of the problems of on/off control, proportional control can be used. Proportional control produces an output which is a multiple of the percentage range of the measured value. The percentage change is determined by the amplifying factor k. The gain of the amplifier adjusts the proportional band of the amplifier.

Output = Gain (k) × Error signal

The disadvantage of proportional control is steady state error, which is the difference between the attained value of the controller and the required value (see Figure 7.25(a)). Steady state error can be reduced by increasing the gain of the controller, but increased gain may cause the controller to become unstable and oscillate. Proportional control is usually used in conjunction with derivative control, or integral control, or a combination of both.

Consider the application of the liquid heating system using steam and a proportional control. The system would be adjusted so that with an input flow of water, the steam valve would be 50 per cent open when the measured value equals the setpoint value. As the temperature reduces or increases, the steam valve will close or open accordingly, in an attempt to maintain the liquid temperature. If the outlet water flow increases, the inlet water flow will increase. As the liquid flow will be greater, the heat loss will

153

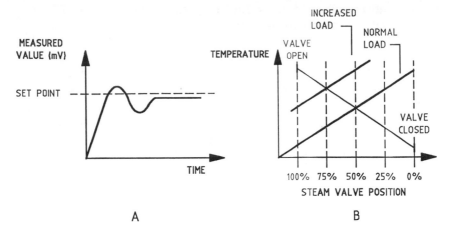

Figure 7.25(a)   Steady state error   (b) Valve position with increased load

increase, so the steam valve must remain more than 50 per cent open to maintain the setpoint temperature. The setpoint is now offset with respect to the valve position, as shown in Figure 7.25(b). With the valve in excess of 50 per cent open, the capacity of the control system to respond to large changes is reduced.

Integral action

The integral action, sometimes termed reset action, responds to the size and time duration of the error signal; therefore the output signal from an integral controller is the mathematical integral of the error. An error signal exists when there is a difference between the process variable and the setpoint, so the integral action will cause the output to change and continue to change until the error no longer exists. Integral action eliminates steady state error. The amount of integral action is measured as minutes per repeat, or repeats per minute, which is the relationship between changes and time.

Derivative action

The derivative mode controller responds to the speed at which the error signal is changing, i.e. the greater the error change the greater the correcting output. The derivative action is measured in terms of time.

Proportional plus integral mode of operation (PI)

A step change in the measurement causes the controller to respond in a proportional manner followed by the integral response, which is added to the proportion response. Because the integral mode determines the output changes as a function of time, the more integral action in the control the faster the output changes.

Proportional integral and derivative control (PID)

The proportion response is the basis for three-mode control (PID). Programmable controllers can either be fitted with input/output assemblies which produce PID control or they will already have sufficient mathematical functions of their own to allow PID control to be carried out. Values such as pressure, flowrate or temperature, can be kept to an amount equivalent to the setpoint by the PID function controlling

154

an output device such as a valve which can be positioned to obtain the desired result. PID control can also be used when the programmable controller function is to position industrial robots. The PID equation is:

PID = Proportional term + Integral term + Derivative term + Bias term

A common PID equation that can be used to obtain PID control is:

$$Co = K\ (\underbrace{E}_{\text{proportional}} + \underbrace{1/Ti\int_{o}^{t}Edt}_{\text{integral}} + \underbrace{KD\ [\ E - E\ (n-1)\ ]\ /\ dt}_{\text{derivative}}) + \text{bias}$$

where Co = control output

K = controller gain (no units)

$1/Ti$ = reset gain constant (repeats per minute)

KD = rate gain constant (minutes)

dt = time between samples (minutes)

bias = output bias

E = error; equal to measured minus set point or setpoint minus measured

$E(n-1)$ = error from last sample.

This PID equation is defined by the international standards association (ISA).

Figure 7.26 shows a PID control valve where a feed forward signal is used to control the position of the outlet valve, and therefore control flow and pressure.

The PID function must have certain characteristics to ensure that the control output behaves in the desired manner. The most important characteristic is deadband. The deadband of the controller is usually a selectable value which determines the error range above and below the setpoint that will not produce an output as long as the process variable is within the set limits. The inclusion of deadband eliminates any hunting by the control device around the setpoint. This occurs when minor adjustments of the controlled position are continually made due to minor fluctuations

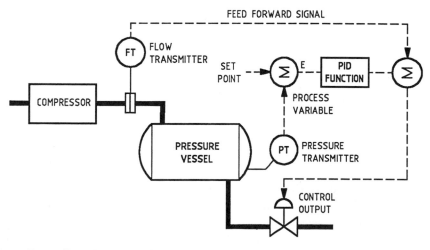

Figure 7.26   Feed forward PID system

in the PID signals. The response of a PID controller to a small change in the process variable is shown in Figure 7.27(a). The process variable will decrease at the occurrence disturbance, and be detected by the PID function which will compensate for the disturbance and return the controlled output to the value determined by the setpoint.

In Figure 7.27(b) the setpoint is changed from a value to a higher value which will cause the PID control to change the controlled output to a greater value.

For further reading on PID control see *Fundamentals of Process Control Theory* by Paul W. Murril (Instrument Society of America).

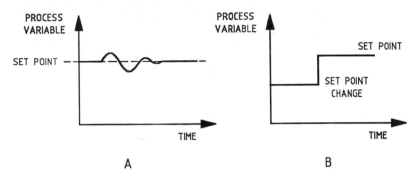

Figure 7.27   Setpoint change and PID control

7.9 Test questions

Multiple choice

1 Which of the following values would be measured by a secondary device?
   (a) Temperature.
   (b) Illumination.
   (c) Force.
   (d) Shaft position.

2 A strain gauge transducer is used to measure:
   (a) force.
   (b) resistance.
   (c) distance.
   (d) temperature.

3 Resolution, with respect to transmitters, can be described as:
   (a) obtaining a similar result.
   (b) the smallest detectable input change.
   (c) how well the instrument measures.
   (d) the ability of the instruments to be correct.

4 The fundamentals of a data acquisition system encompass:

    (a) timing theory.

    (b) quantizing theory.

    (c) sampling theory.

    (d) converter codes.

5 Types of process control can be characterized as:

    (a) open and closed loop.

    (b) closed and voltage loop.

    (c) open and current loop.

    (d) setpoint and current loop.

True or false

6 On/off control eliminates hunting.

7 PID control eliminates steady state error.

8 PID control requires zero deadband.

9 The integral action (termed reset) responds to the size and duration of the error.

10 Closed loop control contains a feedback element.

Written response

11 Discuss the difference between batch and continuous processes.

12 Describe the difference between an analog and a digital input signal.

13 Describe the advantages of a 4 to 20mA current loop as an input signal compared to a 0 to 5V input signal.

14 Discuss the advantages and disadvantages of open loop control compared to closed loop control.

15 Discuss the function of the blocks necessary to make up a data acquisition system.

Data manipulation

8.1 Program editing

All programmable controllers support a range of edit functions to allow changes to programs. The edit functions are capable of being carried out 'on line' or 'off line'. Off line program changes allow:

- functions to be deleted, changed, or inserted;
- rungs to be inserted;
- rungs to be deleted, changed, or inserted;
- words in blocks or word functions to be changed;
- instructions or words to be searched for;
- memory to be cleared.

The edit functions are often carried out while the programmable controller is operating in the program mode. This mode of operation is termed off line program editing. On line program changes can be made to the program while the programmable controller is operating in the run mode. On line program changes must be absolutely accurate and the implications of the changes considered prior to being carried out, as changes on line can cause machinery to operate haphazardly and possibly endanger personnel or damage machinery. On line data changes are often used to modify the data in word or block functions while program troubleshooting is being carried out. Apart from the extra care required, the programming techniques for on line programming are similar to those used for off line programming. Study your programmable controller manual carefully to fully understand the implications of on line program editing before undertaking an edit function.

The programmable controller is provided with help functions which allow access to a series of directories. The directories can be either displayed on the program terminal, or provided as hard copy in the manuals. They allow the user to locate and use the edit functions to make program changes.

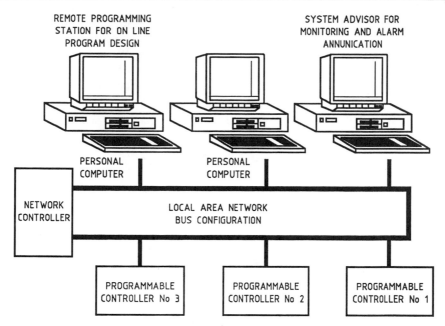

Figure 8.1   Computer programming and advisor PLC system

Some programmable controllers have search functions which allow the user to locate any data table point programmed into a ladder diagram. The search function is useful for locating ladder rungs containing specific functions, or rungs with specific input output addresses. In most PLCs an instruction can be changed by locating the cursor on the instruction and entering the new data.

Many programmable controllers allow individual bits of registers, or the data table, to be monitored, forced on or forced off. The monitoring of an individual bit is useful when troubleshooting, as the status of individual bits can be observed while the program is in operation. The observation allows the user or programmer to locate incorrect bit patterns which inhibit the correct operation of the user program. Forcing individual bits on or off is also used when troubleshooting programs or fault finding in new installations. Forcing a bit on or off may result in outputs becoming active or inactive. When it is necessary to change the status of a bit by forcing, the consequences of such a change must be fully examined and an exact prediction of the result of the change determined before the program forcing operation is allowed to take place. Errors can result in injury to personnel or damage to equipment due to the haphazard operation of a machine or process connected to the PLC. The force instruction can also be used to check the operation of a program by forcing an input that may be operated for a short duration, 1 second or less, and observing the operation.

Computer-based program functions

Programmable controller manufacturers have identified the advantages of using personal, mini, or mainframe computers to allow the program designer to enter, edit, test, and store a program away from the industrial environment and therefore the programmable controller. Manufacturers, as well as software designers, have produced a wide

159

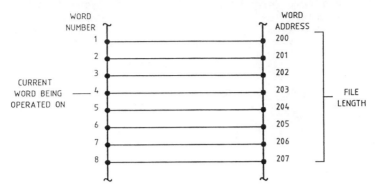

Figure 8.2   File construction

range of software which allow PLC programs to be designed and stored on a computer floppy disk and loaded into the programmable controller when the program designer is satisfied that the program is ready to be tested in the programmable controller.

The computer used for program design also has the facility to provide extensive PLC system documentation. The software used to develop the PLC program will often operate in a personal computer with a data network to connect the personal computer to the programmable controller. If more than one programmable controller is used in the system, a data communication network, sometimes termed a local area network, is used to provide the data communication. A second computer function can be to use the computer as a system advisor or monitor so that a plant control room, located some distance from the industrial environment, can monitor and oversee the process controlled by the programmable controller. Figure 8.1 shows a block diagram of a system where a programmable controller is using both an advisor system for status and alarm annunciation, and a remote programming system via a local area network.

8.2 Word and file moves

The manipulation of entire words is an important feature of a programmable controller. This feature enables PLCs to handle inputs and outputs containing multiple bit configurations such as analog inputs and outputs. Arithmetic functions also require data within the programmable controller to be handled in word format. The concept of words being handled in entirety, and the creation of files in which to store the words will be discussed in this chapter; word applications will be discussed in Chapter 9.

A file is defined as consecutive data table words which have a defined start and end, and are used to store information. The file is scanned as part of the user program with the information on each word being acted upon sequentially. The start address of the file will contain information which will indicate that the following words are part of a single file and how many words are contained within the file. Figure 8.2 shows the construction of a file containing eight words starting at word address 200. Word 204 is shown as the word currently being operated on in the file.

Files must be designed in such a way as not to overlap because overlapping files will cause overwriting of one another, resulting in the file data being corrupted.

160

In some instances it may be necessary to shift complete files from one location to another within the programmable controller memory. Such data shifts are termed file-to-file shifts. File-to-file shifts are used when the data in one file represents a set of conditions which must interact with the programmable controller program a number of times, and therefore must remain intact after each operation. Because the data within this file must also be changed by the program action, a second file is used to handle the data changes and the information within that file is allowed to be altered by the program. The data in the first file, however, remains constant and can therefore be called upon to be used a number of times. Figure 8.3 shows a file-to-file move where the first file, file A, of six words, has the constant data, and file B has the data which is changed.

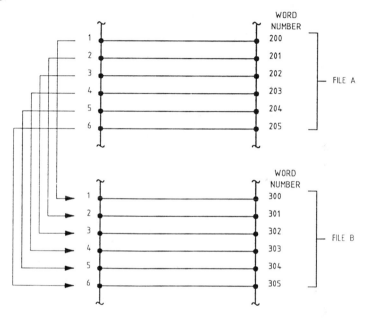

Figure 8.3   File-to-file moves

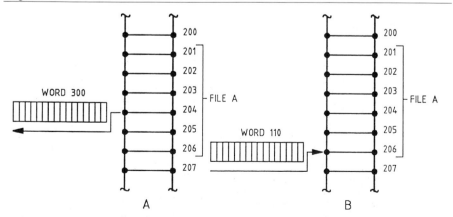

Figure 8.4   Word-to-file and file-to-word moves

Word-to-file and file-to-word moves

There are many instances where it may be necessary to move a single word from a data table location to a file so that the information contained within the word can be manipulated; it is also possible that a word located within a file may be the result of a manipulation and it may be necessary to shift the word from the file to a location in the data table where it can be used as an output. Such moves, involving words and files, are termed word-to-file or file-to-word moves. Figure 8.4(a) shows a file-to-word move where the word at location 204 within file A is transferred to word 300. Figure 8.4(b) shows word 110 transferred into word 206 of file A.

Word-to-word moves

Single word moves to new data table locations are used in programmable controllers for handling arithmetic functions and simple data shifts. An input word from a thumbwheel switch, for example, can be shifted to a word where a seven-segment display is located, therefore causing the setting of the thumbwheel switch to be displayed on the seven-segment display.

8.3 Shift register

Shift register functions can be programmed into programmable controllers to provide sequential control functions. Shift registers can be organized into two types: queues, which provide a first in first out (FIFO) operation, and stacks, which provide a last in first out (LIFO) operation. The shift register can provide storage of bits which occur in a serial sequence, therefore making the shift register useful when approximating the operation of a production line where a specific sequence of operations must occur to produce the desired product. The shift register which approximates the production line is a synchronous type, sometimes termed a serial shift register, where information representing each point on the production line is shifted from one bit to the next, or from one word to the next. The shifting of information can be either from left to right, or right to left.

Figure 8.5 shows the basic operation of the shift register where a common shift pulse or clock causes each bit in the shift register to move one position to the right; this makes the data of the last bit of information fed into the shift register the last in.

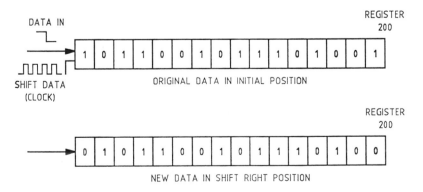

Figure 8.5   16-bit synchronous shift register concept

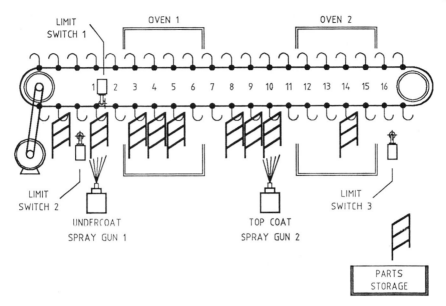

Figure 8.6   Shift register spray painting application

At some point in time the number of data bits fed into the shift register will exceed the register's storage capacity. When storage capacity is exceeded, the first data bits fed into the shift register by the shift pulse are lost out the end of the shift register; this makes the shift register shown in Figure 8.5 a first in first out (FIFO) type. Programmable controller shift registers are often programmed to contain 8, 12 or 16 points which correspond to the programmable controller register size, but it is also possible to produce shift register functions of any number of points by using multiple words.

Figure 8.6 shows a shift register used to control a spray painting operation in a production line. The production line has a number of fundamental requirements:

- painting takes place only when an object to be painted is present on the rack;
- undercoat is applied by spray gun 1;
- undercoat is dried in oven 1;
- top coat is applied by spray gun 2;
- top coat is dried in oven 2;
- all sprayed parts are counted and stored.

The shift register function is used to keep track of the items to be sprayed. As the parts pass along the production line, the shift register bit patterns represent the items on the conveyor hangers to be painted. The logic of this operation is such that when a part to be painted and a part hanger occur together, indicated by the simultaneous operation of LS1 and LS2, a logical 1 is input into the shift register. The logical 1 will cause the undercoat spray gun to operate, and, ten steps later, when a 1 occurs in the shift register, the top coat spray gun is operated. Limit switch 3 counts the parts as they exit the oven. The count obtained by limit switch 1 and limit switch 3 should be equal at the end of the spray painting run and is an indication that the parts commencing the

163

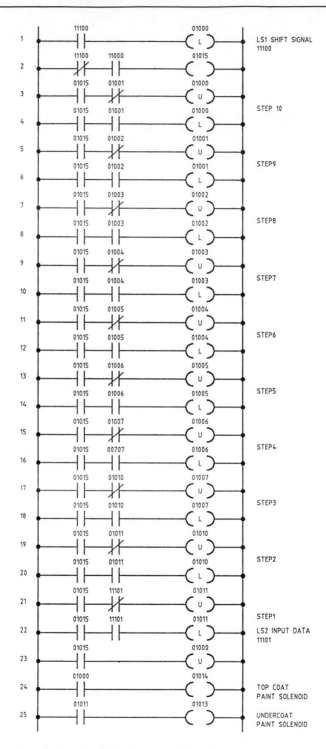

Figure 8.7   Conventional ladder diagram shift register

spray painting run equal the parts that have completed it. A logical 0 in the shift register indicates that that conveyor has no parts on it to be sprayed and therefore inhibits the operation of the spray guns.

Figure 8.7 shows a conventional ladder diagram which produces the effect of a synchronous shift register containing ten steps. The operation of the shift register occurs each time limit switch 1, designated 11100, closes, causing a pulse or clock to be generated to step data through the latches of the shift register. The data, either a 1 or 0, is dependent on the operation of limit switch 2 for a 1, and non operation of limit switch 2 for a 0. A 1 at position 1 or 10 operates the spray solenoid for undercoat or top coat respectively, as shown in rungs 24 and 25.

Most programmable controllers have a shift function, LIFO, or shift register available to the program designer, and in many instances shifting of either words or bits to the left or right is possible.

The Hitachi P 250E has both word and bit shifts, either left or right, available as functions. Figure 8.8(a) shows the ladder diagram of a shift bit left function. Figure 8.8(b) shows the arrangement of the register before and after the shift function execution.

8.4 Last in first out (LIFO)

The LIFO register inverts the order of the data it receives by outputting the last data received first, as the name implies. Another term for the LIFO is the stack. The main

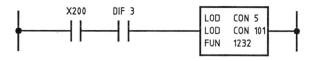

```
X200          DIF 3              LOD   CON 5
                                 LOD   CON 101
                                 FUN   1232
```

X200   - INPUT 200
DIF 3   - EDGE DETECTION
LOD CON 5   - NUMBER OF BITS TO BE SHIFTED (SHIFT COUNT)
LOD CON 101 - START NUMBER OF BITS TO BE SHIFTED
FUN 1232   - SHIFT LEFT FUNCTION

A

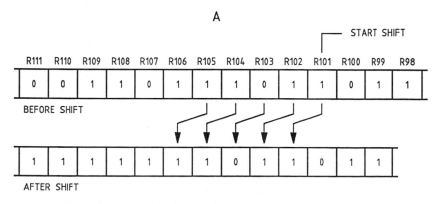

B

Figure 8.8(a) Hitachi P 250E bit left   (b) Bit positions

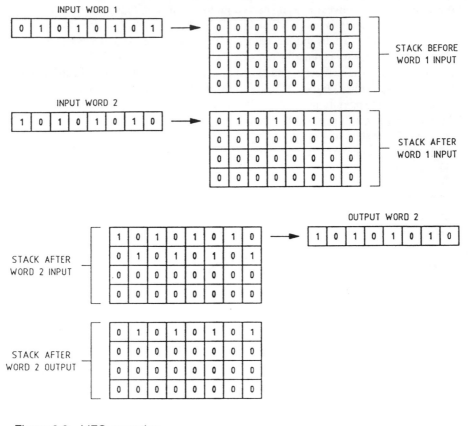

Figure 8.9   LIFO operation

advantage of the LIFO is that information can be added to the stack without disturbing the information that it already contains. Figure 8.9 shows the arrangement of the LIFO where 8-bit words are shifted into the LIFO and taken out as required.

Asynchronous shift register

The asynchronous shift register can compensate for data at the input being fed in at irregular rates, as asynchronous shift registers do not have a regular clock pulse. The asynchronous shift register must have a look-ahead circuit at each step to determine if the next step is ready to receive data.

8.5 Stepladder functions

The stepladder program facility is provided by some programmable controller manufacturers to allow programming of sequential functions directly from a sequential function flow diagram. The advantage of stepladder function is in its programming simplicity, as complex interlocking and control logic is eliminated.

8.6 Test questions

Multiple choice

1 Off line programming can be carried out when the programmable controller is in the:
   (a) run mode.
   (b) program mode.
   (c) operate mode.
   (d) test mode.

2 One advantage of using a personal computer as a programmable controller is:
   (a) the mass storage capacity of the computer floppy disc.
   (b) the computer is faster.
   (c) the computer cannot operate in an industrial environment.
   (d) the computer cannot be connected to the programmable controller.

3 Which of the following would not be described as a shift register?
   (a) FIFO.
   (b) LIFO.
   (c) Stack.
   (d) Register.

4 A practical application for a shift register is:
   (a) data manipulation.
   (b) arithmetic functions.
   (c) word-to-file transfers.
   (d) production line simulation.

5 The LIFO is sometimes termed a:
   (a) file.
   (b) stepladder.
   (c) FIFO.
   (d) stack.

True or false

6 The term asynchronous indicates the shift register is not synchronized with the clock.

7 On line programming can be carried out while the program is operating.

8 Forcing a 1 into a data table bit will result in all the 1s in the data table word becoming 0.

9 Computers can be used for programmable controller program edit and design.

10 A local area network interconnects programmable controllers and computers using data communication techniques.

Written response

11 Describe the possible consequences of incorrect program changes when on-line program alterations are carried out.

12 Describe the construction of a file.

13 Describe the operation of a file-to-file move.

14 Design a ladder diagram for the spray paint booth shown in Figure 8.6 which will allow total parts in and total parts out to be compared.

Advanced programming techniques

9.1 File construction and manipulation

A file is a group of consecutive words, often in a special location in the programmable controller memory, which can be used to store and manipulate information which would have some relevance to other information in the file. A typical file construction is shown in Figure 9.1 where ten words go to make up the file from address 400 to 411. In this case the ten data table words are counted in octal.

The data in a file can be changed by using a word-to-file move or shifted to another file by using a file-to-file move. Shifting data between files allows the data in the original file to be retained while manipulation of the second file takes place. The transfer of information from source file A to receiving file B can be achieved through one of the following methods:

- a complete transfer, where the entire contents of file A are transferred to file B during one scan. A complete file transfer per scan slows down the programmable controller processor which will increase the scan time and therefore the program execution time;

- a distributed complete transfer, where a predetermined number of words of file A are transferred to file B each scan, therefore taking several scans to transfer the information between files;

- an incremental transfer, where the individual words of file A are transferred to file B only when a true/false condition is set in a ladder diagram rung.

	WORD ADDRESS	
WORD 1		400
WORD 2		401
WORD 3		402
WORD 4		403
WORD 5		404
WORD 6		405
WORD 7		406
WORD 8		407
WORD 9		410
WORD 10		411

FILE A

Figure 9.1   File construction

9.2 Word input and output devices

Consider a liquid storage vessel that has upper and lower level limit switches and a valve actuator to control the inlet liquid as shown in Figure 9.2. In this example, the temperature of the liquid is to be kept between 95°F and 105°F; an alarm will be triggered should the temperature of the liquid fall below 97°F, or be higher than 102°F. A thumbwheel switch is used to set the temperature setpoints and a seven-segment LED display is used to display the setpoint value. When the setpoint switch is activated, the setpoint value is displayed on the seven-segment display, indicating to the operator that the temperature of the liquid is out of limits (see Figure 9.3).

Rungs 1 and 2 in the program are a start/stop circuit. This circuit also contains an inhibit (using the setpoint enable switches) to stop the system when the setpoints are being set. Rungs 3 and 4 open and close the liquid inlet valve; rungs 5 to 8 allow the

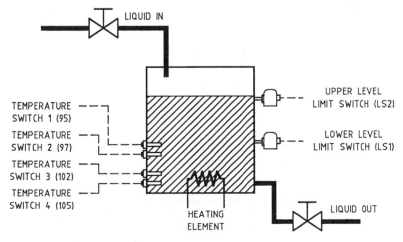

Figure 9.2   Liquid control application

data at the thumbwheel address to be placed into consecutive words 400 to 403. Rung 9 allows the file consisting of words 400 to 403 to be transferred to a file at words 500 to 503 when the system is running. If the setpoints are being set, the file-to-file move will not take place. Rungs 10 to 13 allow the setpoint value to be displayed while it is being set up. Rungs 14 to 17 cause alarms when the temperature switches are operated, i.e. when the temperature is out of limits. Rung 18 closes the liquid outlet valve, and rungs 19 to 20 cause the setpoint value to be displayed, but only when the system is running and the temperature-limit switch is operated. This ladder diagram program is suitable for an Allen-Bradley PLC 2 series programmable controller.

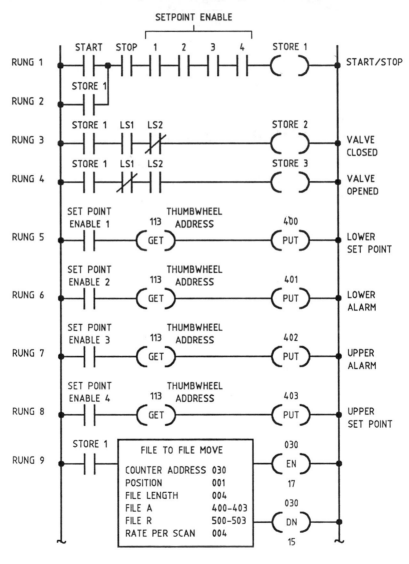

CONTINUED ON NEXT PAGE

Figure 9.3   File and word manipulations

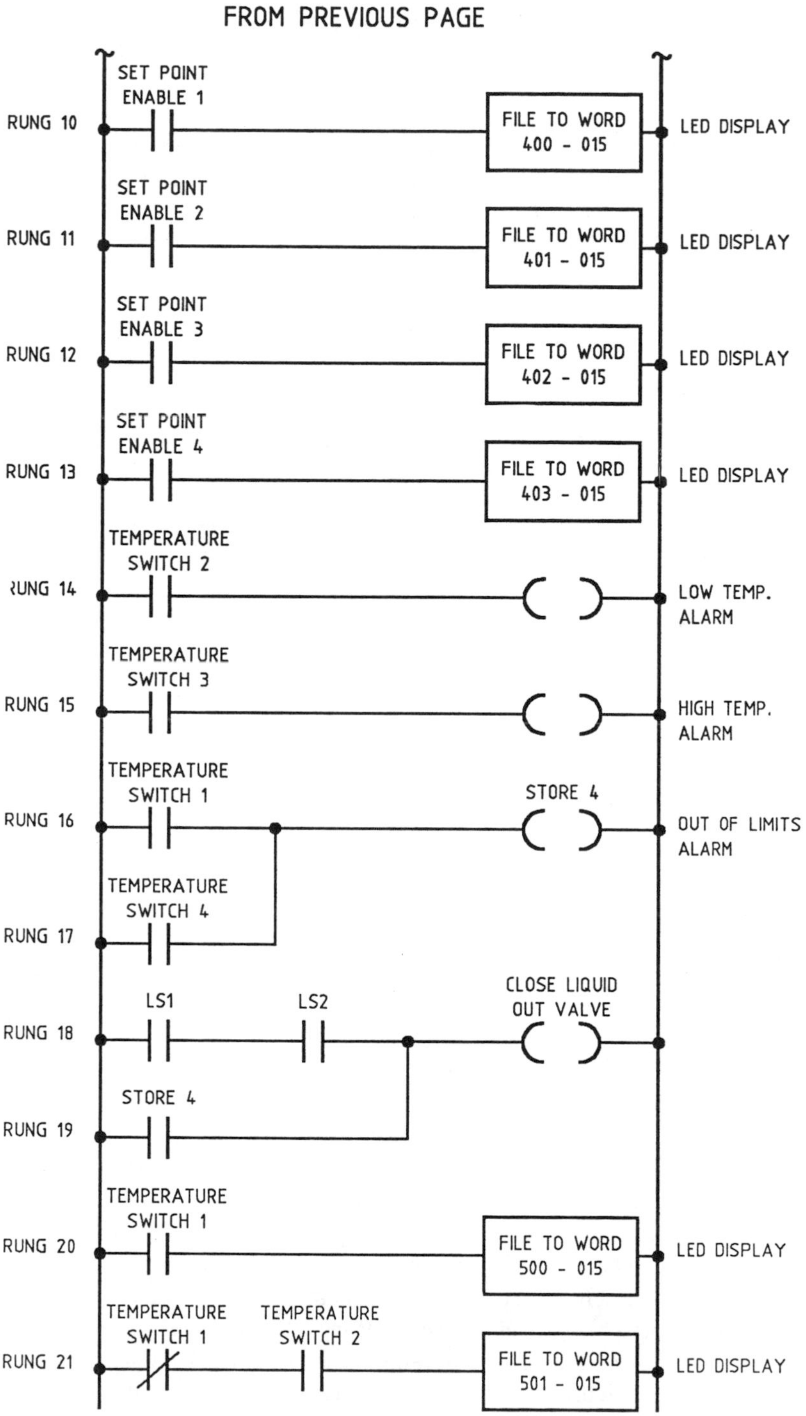

Figure 9.3 File and word manipulations

The creation of files to allow data from word input devices to be directly located into a file makes programming easier, allows a more efficient use of the data provided by the word input device, and reduces ladder diagram rungs. Figure 9.4 shows a ladder diagram where data from a thumbwheel switch is input to a file when input 11000 is true in rung 1. Rung 2 subtracts a constant value. Rung 3 outputs the result to a seven-segment display. The thumbwheel switch is a word input device; the seven-segment display is a word output device. This program is suitable for an Allen-Bradley PLC 2 series programmable controller.

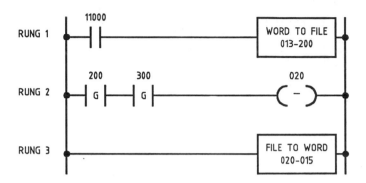

Figure 9.4   Word input and output devices

9.3 Analog input and output signals

In complex control applications where programmable controllers are used, such as PID control, servo motor control, robot system sequencing, and position control, it is necessary to provide inputs which contain data bits in excess of one word. Word input assemblies are often termed intelligent input/output devices. Many will have a microprocessor on board to enable the complex function to be carried out. The data bits are presented to the programmable controller as an analog signal which is converted to the digital equivalent of the analog.

The analog information is read into the programmable controller as a block of digital data after the analog-to-digital conversion has taken place. The block of data is placed into a file where it can be manipulated. The file information can then be written as a block to an output assembly where, after conversion to the analog equivalent of the digital information, it can be output as an analog signal. A block read and block write function, used together on the same input/output assembly to input an analog value and output an analog value, is termed a bi-directional block transfer. Figure 9.5 shows a block diagram of the programmable controller program where an analog signal is input using a block read and output using a block write.

Figure 9.6 shows a ladder diagram suitable for an Allen-Bradley PLC 2 programmable controller that uses a block transfer read to input an analog into a file consisting of six words starting at word address 0220. When the rung is true, the block transfer read is enabled, and the enable bit becomes true. When the block has been read, the

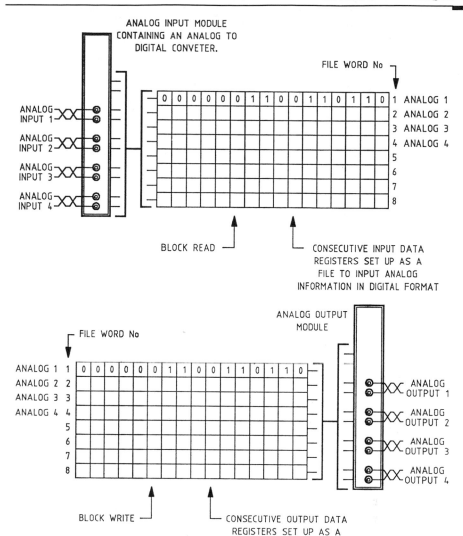

Figure 9.5   Block read/write format

done bit is true. The block transfer read and write are defined by using a counter/timer address. The analog input module is located at address 160 which indicates:

rack   1
module group 6
slot   0

The analog output module is located at:

rack   1
module group 7
slot   0

In this application, the rack is configured for single slot addressing.

173

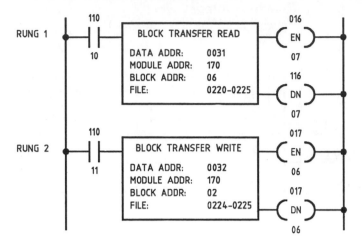

Figure 9.6   Analog input and out put using block transfers

The block transfer write command reads the analog value out of the file, which is a two-word block, starting at word 224. The file is written to module address 170, which is the location of the analog output module when rung 2 is true.

In programmable controller analog applications where analog input and output are used, the block transfer read and write are preceded by condition logic that makes the read and write functions automatic, utilizing the enable and done bits. Run-time errors can be encountered if the block transfer read and write are not enabled and completed in the correct sequence. The analog input/output modules have a number of scaling words which allow the programmable controller to set up the module for the required application. The manufacturer's handbook explains the scaling of analog modules in detail.

To set up an analog assembly, the unit must be configured to the particular system requirements. The system requirements must be assessed and either programmed into the module or set by hardwired switches prior to installing the assembly into the system. The scaling of the assembly, which is part of the assembly configuration, allows unscaled data such as degrees centigrade, pounds per square inch, and liters per minute to be converted into engineering units.

An example of unscaled data may be an analog which produces a binary value of 0 – 4095 for a 0 – 100°C temperature. To read the value conveniently, the output can be sealed so that 0 binary displays 0°C, and 4095 binary displays 100°C.

9.4 Sequencers

The sequencer produces output data in a sequential order and is used to control machines which operate in a sequential manner. The sequencer operates one step at a time, each step being completed prior to producing the outputs required for the next step.

The electric drum sequencer, where an electric motor turns a drum at a specified speed and pegs cause cams to operate contacts located along the drum's length, has

174

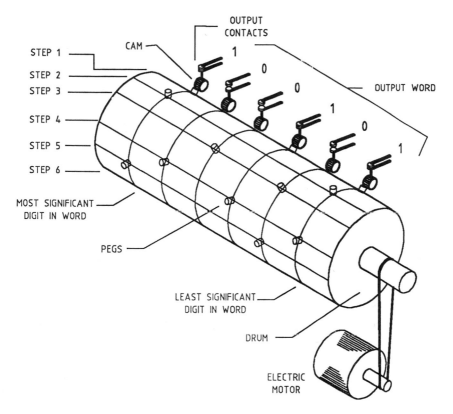

Figure 9.7   Drum sequencer

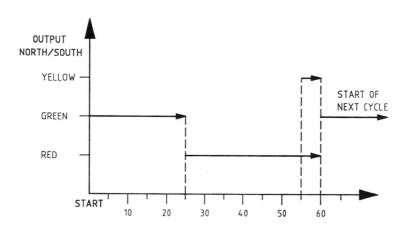

Figure 9.8   Traffic light sequence control

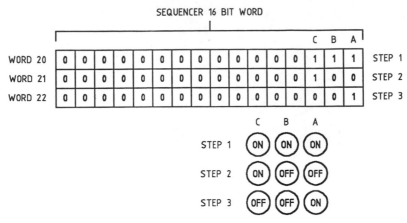

Figure 9.9   Three word sequence

essentially been replaced by the electronic sequencer. Each time the contact is operated, a logical 1 or on condition is produced as an output, and, when a contact is not operated, a logical 0 or off condition is the result. Figure 9.7 shows the basic construction of a drum sequencer. The program for the drum sequencer is determined by the location of the pegs; each row of pegs represents the steps in the sequencer program. The output of the program occurs when the pegs cause the contacts to open or close.

The operation of a sequencer provides a direct relationship between an action and time, therefore a timing diagram can be drawn for it. Figure 9.8 shows the timing sequence diagram of a traffic light which has PLC sequence control. The use of a sequencer is an advantage in any application where a regular occurrence of events is part of the process.

The programming of sequences will vary between programmable controller manufacturers, but the operational concepts are the same. The sequence of events controlled by the sequencer is determined by the bit pattern of each consecutive word and the number of words in the sequence. In Figure 9.9, for example, a three-word sequence of events is controlled. At the first step in the sequence, outputs A, B, and C are on. At the second step, only output C is on, and at the third step output A is on. The sequence, once complete, will be restarted on the next processor scan and will continue to produce the same output pattern in a cyclic manner. Each time a 1 is in the word and the processor is scanning that word, the outputs associated with the word become active.

A mask is used in conjunction with the sequencer to control which outputs are allowed to become active at any given sequence step. This is useful in applications where it is necessary to selectively stop word data from appearing in the output word. Figure 9.10 shows how the mask operates. If a 1 has been set in the mask, the output is allowed to become active by allowing data transfer to the output file; if a 0 is set in the mask, data is prevented from being transferred to the output file.

Output word 20, shown in broken lines, is the expected output as a result of the mask and the data in step 2. The transition to step 3, the current step, and retention of the mask word bit pattern results in the output bit pattern of word 20, shown in solid lines.

	17	16	15	14	13	12	11	10	07	06	05	04	03	02	01	00	
WORD 20	0	0	0	0	1	1	1	1	0	0	0	0	0	1	0	1	OUTPUT WORD STEP 2
WORD 20	0	0	0	0	0	0	0	0	1	1	1	1	0	0	0	0	OUTPUT WORD STEP 3
WORD 25	0	0	0	0	1	1	1	1	1	1	1	1	0	1	0	1	MASK WORD

	17	16	15	14	13	12	11	10	07	06	05	04	03	02	01	00	
WORD 26	0	0	0	0	0	0	0	0	0	0	1	1	0	1	0	1	STEP 1
WORD 27	0	0	0	0	1	1	1	1	0	0	0	0	1	1	1	1	STEP 2
→ WORD 28	1	1	1	1	0	0	0	0	1	1	1	1	0	0	0	0	STEP 3
WORD 29	1	0	1	0	0	0	0	0	1	0	0	0	1	0	0	1	STEP 4
WORD 30	0	0	0	0	0	0	0	0	0	0	0	0	1	1	1	0	STEP 5
WORD 31	0	1	0	0	0	0	0	0	0	1	0	1	0	1	0	0	STEP 6

CURRENT STEP
IN SEQUENCE

Figure 9.10   Mask operation

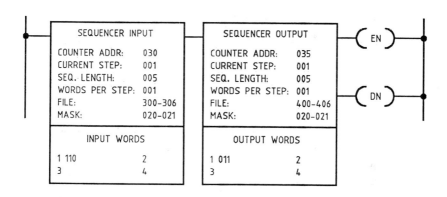

Figure 9.11   Sequencer program rung

Note that the only data bits in the output word that change are those where the corresponding bit in the mask is a 1, i.e. bits 04 to 07, and 10 to 13.

The output bit pattern for the bits where the mask is 0 remains unchanged. By using a file-to-word or word-to-word shift, it is possible to rearrange the bit pattern of the mask which can be changed to suit the application.

The configuration of the data for a sequencer is similar to that of a file, in fact the sequencer data is stored as a pattern of bits in a file. To program a sequencer into an Allen-Bradley PLC 2 programmable controller sequencer, an input file is created which holds the bit pattern of the sequence read from a number of input words, and an output file is used to output the sequence to the required devices. The bit pattern of the mask word controls which outputs are made active. Figure 9.11 shows a ladder diagram rung with the sequencer input and output file.

9.5 Sequence controllers

The stand alone sequence controller is a close relation to the programmable controller; it is considered by some to be a type of dedicated programmable controller. The sequence controller considered will be the Omron Sysmac-Po unit which is a stop/advance sequence controller. Omron also manufacture a wide range of programmable controllers. The applications of sequence controllers are similar to programmable controller applications and include the control of machinery such as multiple robot sequences, conveyors, and sorting functions, as well as process control applications, and repetitive testing equipment.

The program to control the sequence of events is entered via a key pad into the memory of a microprocessor-controlled device. Outputs are turned on or off under the control of the program. The instructions that are available with this sequencer are shown in Table 9.1; the block diagram of the sequencer controller is shown in Figure 9.12.

Table 9.1   Sequencer instructions

Operation code	Title
NOP	Next output
END	End
RET	Return
JMP	Jump
CJP	Conditional Jump
AND	Logical AND
OR	Logical OR
CNT	Counter
TIM	Timer
TMC	Interval timer check
RPT	Repeat
FSR	Flag set
FSR IN	Flag reset

The facilities of the Omron Sysmac-Po include:
- a program capacity of sixty-four steps, which includes two steps for interrupt processing;
- timing chart programming system;
- twelve input points;
- twelve output points;
- a counter function for between 1 and 99 counts;
- a timer with three time-setting ranges of:
  1 0 hr 00 min to 9 hr 59 min
  2 0 min 00 sec to 59 min 59 sec
  3 00.0 sec to 59.9 sec.

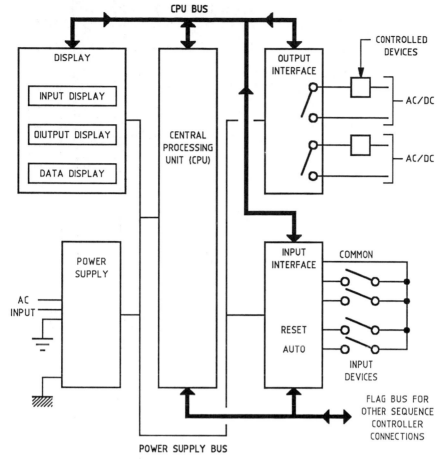

Figure 9.12   Sequencer block diagram

The sequence controller uses step advance programming to produce a sequence of outputs depending on the step advance conditions. The step advance condition is the instruction logic. The logical steps required to produce a sequence controller program are similar to those which should be used when programming a programmable controller:

1 Prepare a sequence chart for timing purposes.
2 Assign input/output terminal connections.
3 List the program steps on a program chart.
4 Write and run the program.
5 Edit the program and re-test if necessary.
6 Commission the operating system.

To program the sequencer, the instructions and program steps are programmed into the sequencer's memory from the program console. Figure 9.13 shows the program console of the Omron Sysmac-Po sequence controller with a summarized list of the key functions and abbreviations.

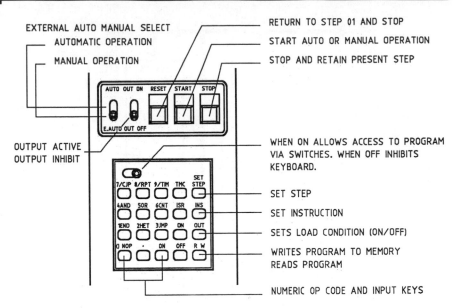

Figure 9.13   Omron-Sysmac Po program console

Before commencing programming it is necessary to clear the memory of the sequencer. The previous contents of the memory will have been retained because power is maintained by the back–up battery. The keystrokes necessary to clear the memory are:

1 switch out on 'on';
2 switch console switch 'on';
3 operate SET STEP;
4 operate 0/NOP;
5 operate 0/NOP;
6 operate INS;   } operated together
7 operate R/W;
8 operate reset.

When the memory is clear, the display will flicker and indicate as shown in Figure 9.14.

To write a program:

1 console switches 'on';
2 set step followed by the instruction key:

step	01	02	03	
op code	4	5	4	
data 1	08	02	04	} input logic
data 2	06	04	08	
	R/W	R/W	R/W	
	OUT	OUT	OUT	
	ON	OFF	ON	
	OFF etc	ON etc	ON etc	

180

STEP   OP   DATA 1   DATA 2

Figure 9.14   Memory clear display

The program can be run when the console switch and the START button is pressed. The sequencer must be in AUTO and the OUT ON key operated to cause the outputs to operate. The input logic must be correct or the sequencer will not step, i.e. for the first step input 08 AND 06 must be operated, for step 2 input 02 or 04 must be operated, and so on. The output functions are set during the programming of a step. The OUT button, followed by the required ON/OFF configuration, sets the outputs.

Applications for sequence controllers include conveyor forward/reverse control, machine tool control, simple robot sequencing, and chemical manufacture mixing and heating processes. If the sequencer is being used for a complex operation a program chart may be required to assist in program design. Figure 9.15 shows the information required for a program chart.

9.6 Multiplexing inputs and outputs

In many PLC applications the programmable controller will have a large number of inputs or outputs, or both inputs and outputs. If, say, three banks of BCD thumbwheel switches, each bank containing four digits, and a PLC input point is provided for each BCD bit, forty-eight individual input points are required. If the inputs are multiplexed, the number of input points can be substantially reduced. Figure 9.16 shows eight input switches in BCD format multiplexed into four inputs with the input wiring configuration required for the multiplexed inputs. As the output enables each group of four input switches, the input switch data is read into the input module. One output is enabled at a time, once each scan, thus producing time division multiplexing. Additional

STEP	OP	DATA1	DATA2	OUTPUT SETTINGS											
				1	2	3	4	5	6	7	8	9	10	11	12
00															
01	4	08	06	ON	OFF	ON	OFF	ON	OFF	ON	ON	OFF	OFF	OFF	OFF
02	5	02	04	OFF	ON	ON	OFF	OFF	OFF	ON	OFF	OFF	ON	OFF	OFF
03	4	04	08	ON	ON	OFF	OFF	OFF	OFF	ON	OFF	OFF	OFF	ON	ON
04															
05															
06															
07															

Figure 9.15   Sequencer program chart

181

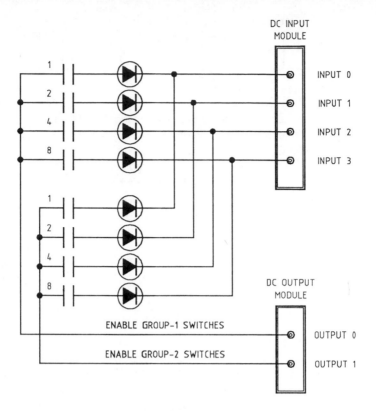

Figure 9.16   Multiplexing inputs

groups of input switches can be connected using a separate output to enable each group. The diodes shown in Figure 9.16 are required to stop back feed between inputs of each group. The disadvantage of multiplexing inputs is that input isolation between controls, in this case the input switches, is forfeit. The reduced requirement for input points and therefore modules, however, is often a more important consideration.

The ladder diagram program to multiplex the two BCD inputs in Figure 9.16 is shown in Figure 9.17. The operation of this program is:

1 Rung 1 contains a self-resetting timer which produces the timing pulses necessary to produce the multiplexing function.
2 Rung 2 will produce an output 02000 when timer bit 03000 is set and bits 03001 and 03002 are not set; i.e. the timer time-period one, 02000, will unlatch output 01001, the enable for the second group of BCD inputs (rung 10), and latch output 01000, the enable for the first group of BCD inputs (rung 11).
3 Rung 3 will produce an output 02001 when timer bit 03001 is true and bits 03000 and 03002 are not, which is at time-period two, 02001 gets the input data and puts the data into a storage location at 025, rung 8.
4 Rung 4 represents time-period three and allows for input data settling time.
5 Rung 5 is the time-period four which disables output 0 and enables output 1 so that the second BCD data can be presented to the input rungs 12 and 13 respectively.

182

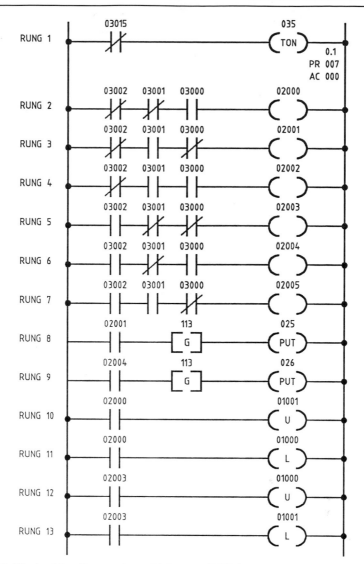

Figure 9.17   Ladder diagram to multiplex two BCD inputs

6 Rung 6 is for time-period five to allow the second BCD data to be input and stored at location 026 in rung 9.

7 Rung 7 allows settling and input data reading time.

8 The seventh timer count resets the counter to allow the procedure to be repeated.

Multiplexed outputs

Multiplexed outputs, particularly when seven-segment displays are used, reduce the number of output assemblies required by using four outputs to provide signals for the display, and an enable line, termed a strobe, to turn on each display at the appropriate time. Figure 9.18 shows the output assembly wiring required when multiplexing two three-digit seven-segment displays.

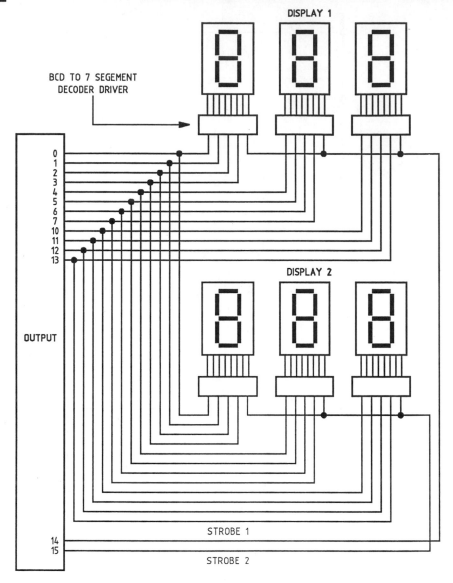

Figure 9.18   Multiplexing seven-segment displays

The ladder diagram program to multiplex two seven-segment displays which accept a BCD input is shown in Figure 9.19. The operation of the ladder diagram suitable for an Allen-Bradley PLC 2 shown in Figure 9.19 is as follows:

1 Rung 1 is a self re-setting timer which produces output pulses used to generate the multiplexing function.
2 Rung 2 is the first time period. In conjunction with rung 8 value at location 110, it is output to location 015.
3 Rung 3, in conjunction with rung 10, applies the strobe to display 1. The strobe is removed at the next increment.

184

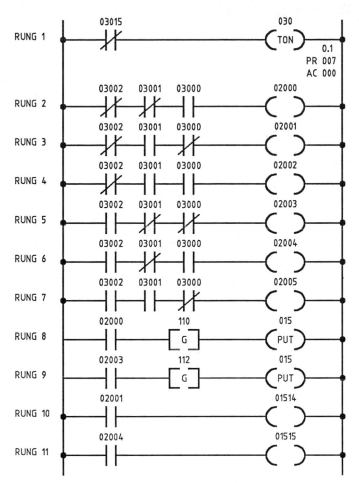

Figure 9.19   Ladder diagram to multiplex two BCD inputs to seven-segment displays

4 Rung 4 allows settling time.

5 Rung 5 inputs the data at address 112 to output 015 for the second display via rung 9.

6 Rung 6 enables the strobe for the second display in rung 11. The strobe for the second display is removed at the timer's next increment.

7 Rung 7 allows settling time.

The display information is updated at the completion of each scan made by the programmable controller.

9.7 Test questions

Multiple choice

1 Files in programmable controllers are considered to be a group of consecutive:
   (a) 16 bits.
   (b) words.
   (c) timers.
   (d) shift registers.

2 Multiplexing input and output:
   (a) makes the system smaller.
   (b) cannot be achieved with less than 128 input/output.
   (c) allows the connection of additional input and output points to the PLC.
   (d) increases the number of input modules.

3 A mask is used when programming:
   (a) words.
   (b) files.
   (c) shift registers.
   (d) sequencers.

4 The disadvantage of multiplexing inputs and outputs is that:
   (a) diodes are required to stop back feed.
   (b) more wiring is necessary.
   (c) more input modules are required.
   (d) isolation between inputs is forfeit.

5 In a step advance sequencer, what condition must the input logic be before a step takes place?
   (a) Automatic.
   (b) True.
   (c) False.
   (d) Reset.

True or false

6 Word input devices allow analog to be fed into the processor memory as a file after A/D conversion.

7 Sequencer outputs occur one at a time, one after another.

8 The configuration of file data in a PLC is similar to the data configuration in a PLC sequencer.

9 PLC analog modules cannot be scaled.

10 BCD thumbwheel switches are word input devices.

Written response

11 Describe the function of a mask.

12 List the steps required to design a sequence controller program.

13 Draw a PLC ladder diagram suitable for multiplexing a seven-segment display.

14 Draw the ladder diagram of a sequence controller.

15 Describe the operation of a drum sequencer.

Data communications

10.1 Introduction

Data communication can be defined as the means of transferring digital information in the form of 1s and 0s from a sender to a receiver. The programmable controller is able to communicate with a variety of devices using data communication techniques. The earliest programmable controller was able to receive from and send to audio cassette recorders as well as to document program information using printers or teletype machines. Later generations of programmable controllers are able to communicate with other types of microprocessor-based devices, such as computer numerical control (CNC) machines, computer-based process control and robot controllers. The programmable controller is often not the central or focal point for the information transfer, instead the computer will usually control a variety of complex operations and machines in a computer-integrated manufacturing (CIM) application.

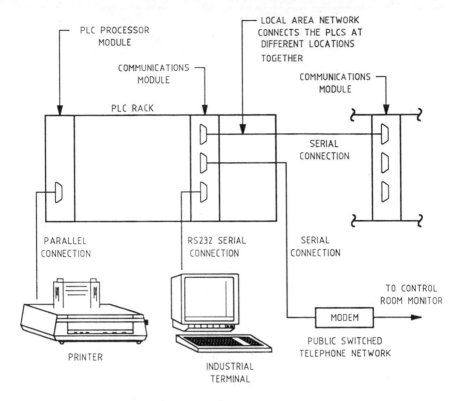

Figure 10.1   Data communications and PLCs

To facilitate the interconnection of data communicating devices, programmable controllers are fitted with a wide range of data communicating input and output terminals, termed ports. Generally the data input and output ports are classified into two groups: serial, and parallel. The terms serial and parallel are a means of classifying the communications channel by the method used to transfer the data from one point to another. The programmable controller may be equipped to facilitate a variety of communication methods, as shown in Figure 10.1. Here the programmable controller is interconnected to a printer via a parallel connection, to a local area network via a serial connection, to an industrial terminal via an RS232c serial connection, and to a control room monitor located some distance away via the public telephone network using serial communications techniques.

10.2 Data transmission speed

The number of data bits that can be transferred between data communicating devices without error and converted to useful information is the maximum speed of a data communicating system. There are a number of factors outside the control of the communicating device which govern the speed of transmission, such as data channel noise, data channel interference, and data channel bandwidth.

As it is desirable to transfer data from one point to another at maximum speed to obtain maximum efficiency, the data channel considerations are important when designing data communication networks.

Data channel noise and interference are unknown quantities which can vary in intensity over the transmission time and result in errors occurring in the received information. Techniques in error detection and correction are available, but such techniques reduce the rate at which information can be transferred between the transmitter and the receiver.

The channel can consist of one of a variety of physical mediums used to interconnect data communication devices. The most common channel mediums are copper conductors, twisted wire pairs, co-axial cable radio bearer, and optical fibers. If the distance between the communicating devices is short, the choice of medium is relatively unimportant as the data circuit losses are minimum. The transmission medium is usually governed by cost and the required data transmission speed.

The bandwidth (usable frequency response) of any channel is determined by the physical properties of the channel. Figure 10.2 shows the equivalent circuit of a two-wire balanced transmission medium which could be used for data transmission and reception. The most significant component of the physical properties of a copper conductor pair is the shunt capacitance which limits the frequency of operation of the transmitted data because it shunts the higher frequency signals more than the lower frequency signals.

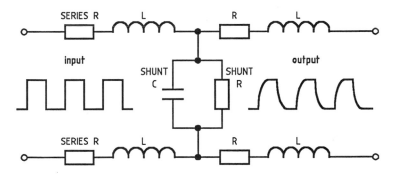

Figure 10.2   Equivalent of a two-wire balanced transmission line

Consider a data transmission consisting of 1s and 0s which are generated in the form of square waves and transmitted directly onto the transmission medium. This form of data transmission is termed baseband. The receiver has to detect the level of transition between a 1 and a 0 to decode the receive data correctly. As the 1s and 0s are produced in the form of square waves and contain all the odd harmonics (multiples of the fundamental frequency), the bandwidth of the channel must be large or the transmitted data signal will be distorted and the possibility of errors in the transmitted information will be increased.

Any desired transmission rates can be achieved in a data transmission system, but for ease of communication between devices a number of standard speeds are used. These are 50, 110, 300, 600, 1200, 2400, 4800, 9600, and 19200 bits of data per second.

In a serial channel, carrying two levels of binary data, there is a relationship between bits of data per second, the baud rate, and the channel bandwidth. The baud rate is the rate at which data is transferred between the transmitter and the receiver. It is independent of the number of levels used in the data transmission and reception process, i.e. of the number of signalling events per second. Multiple level data transfer is used to increase the speed of the data transmission for a given bandwidth. The bandwidth required for a two level system is:

Bandwidth (Hz) = 0.5 × Data rate (in bits per second)

The bandwidth for a multilevel system can be determined by:

$$\text{Baud rate} = \frac{\text{Number of coincident bits}}{\text{Total data rate (in bits per second)}}$$

Bandwidth (Hz) = 0.5 × Baud rate

10.3 Parallel communications

Parallel data communication is satisfactory over short distances of approximately 10 meters (30 feet). Signal losses are low, and high speed is attainable. Parallel communication is therefore often used as a communications technique between a programmable controller and a printer to provide PLC system documentation.

The parallel communication signal is shown in Figure 10.3 where the PLC is communicating with a printer. The parallel connection requires at least nine wires: eight for data bits, and a signal common. The data is fed onto the cable one word at a time, making the signal serial by word with 8 data bits being transmitted simultaneously. Speeds of 9600 bits per second are easily obtainable over short distances using parallel communication.

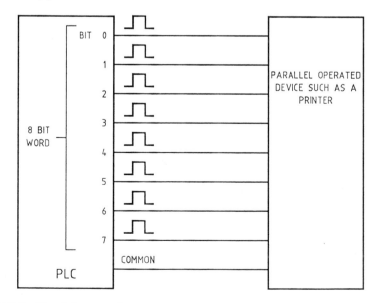

Figure 10.3   Parallel connection

10.4 Serial communications

The serial connection for data communication requires only two wires. With serial communications the data word is transmitted one bit at a time until the word has been sent, then the bits of the next word are sent one bit at a time. The serial connection with a single pair of wires as a medium, as shown in Figure 10.4, has a bandwidth of 3000 Hz and a theoretical maximum data speed of 6000 bits per second. In practice, however, the channel introduces noise and distortion which limits the channel data speed to approximately 1200 bits per second, unless special encoding and decoding techniques are used. Serial communication can be transmitted over greater distances than parallel communication.

Each data word in the serial transmission must be denoted with a known start followed by the data bits which contain the intelligence of the data transmission and a stop bit. A 1 occurring in the data transmission is termed the mark, and a 0, the space.

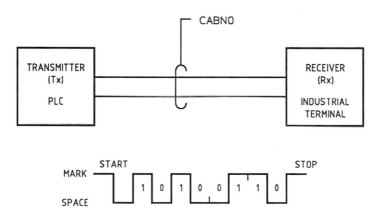

Figure 10.4   Serial transmission

10.5 Simplex, half duplex and duplex

Simplex data transmission occurs in one direction between a data transmitter and receiver and has no facility to indicate that the transmitted signal has been received. Simplex operation is shown in Figure 10.5(a).

Half duplex transmission, as shown in Figure 10.5(b), requires only one pair of wires, but there is a switching arrangement whereby information is passed between the ends in one direction at a time. The information being passed between the two ends allows handshaking to take place. (Handshaking is a predetermined code used to establish the data connection.)

The full duplex operation allows the transmission of data in both directions simultaneously, as shown in Figure 10.5(c). Full duplex operation may require the use of two pairs of wires as the transmission medium.

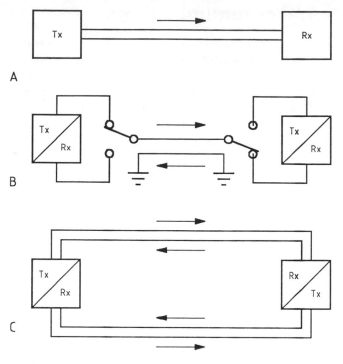

Figure 10.5(a) Simplex   (b) Half duplex   (c) Full duplex

10.6 Asynchronous and synchronous transmissions

There are two common methods used to determine which is the first bit of a transmitted data word. These methods are termed asynchronous and synchronous. The asynchronous transmission is used when one data character is transmitted at a time and the characteristic is framed by a known start condition (prior to the character being transmitted) and a known stop condition (at the end of the character). Between the start and stop condition a number of data bits are located. The number of data bits representing a character is dependent on the code for information interchange that is used. Table 10.1 lists the common codes and the number of data bits to represent a character.

The synchronous transmission differs significantly from the asynchronous transmission in that it carries the clock along with it to lock the transmitter and receiver together. Synchronous transmission transfers continuous streams of data at higher data rates than asynchronous transmission, but the transmitting and receiving circuitry is more complex. Figure 10.6 shows an asynchronous and synchronous data transmission bit pattern. The asynchronous signal, when idle, is a logical 1. The signal going low indicates that a character is about to be sent. The required number of data bits are sent, followed by a known stop condition which, in this case, is two consecutive stop bits. The system is now ready to transmit another character. The synchronous

Table 10.1   Interchange codes and character data bits

Code	Number of data bits per character
Baudot	5
Trans code	6
American standard code for information in interchange (ASCII)	7
Extended binary coded decimal interchange code (EBCDIC)	8

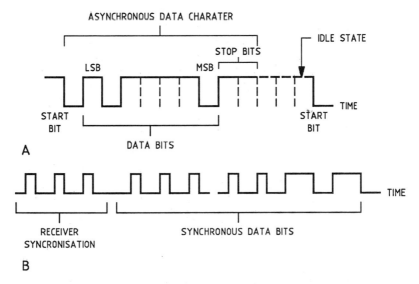

Figure 10.6(a) Asynchronous transmission   (b) Synchronous transmission

transmission commences with a number of clock pulses to initiate the receiver synchronization. After receiver synchronization, the data can be transmitted in a continual stream of bits.

10.7 Codes for information interchange

Three codes used for information interchange that are in common use are Baudot, ASCII, and EBCDIC. The 5-bit Baudot code, named after a French engineer who worked on telegraphy, is known as the CCITT alphabet number 2 and is used internationally on the telegraph (Telex) network. Two forms of Baudot code are used in the USA, the USA commercial form, and the USA AT and T form. The Baudot code is used in conjunction with simple terminals and teletypewriters with limited character sets. Code conversion is often required to operate Baudot code in conjunction with a computer. Table 10.2 lists the CCITT Baudot code. CCITT organization is discussed in the section on data communication interfaces in this chapter.

193

Table 10.2   Baudot code

11000	A	–
10011	B	?
01110	C	:
10010	D	who are you
10000	E	3
10110	F	
01011	G	
00101	H	
01100	I	8
11010	J	BELL
11110	K	(
01001	L	)
00111	M	.
00110	N	,
00011	O	9
01101	P	0
11101	Q	1
01010	R	4
10100	S	!
00001	T	5
11100	U	7
01111	V	=
11001	W	2
10111	X	/
10101	Y	6
10001	Z	+
00000	BLANK	
11111	LETTER SHIFT	
11011	FIGURE SHIFT	
00100	SPACE	
00010	CARRIAGE RETURN	
01000	LINE FEED	

The American standard code for information interchange (ASCII) is a 7-bit code used for digital communications. The 7 bits per character allow 128 bit combinations, which makes the ASCII code suitable for encoding upper and lower case alphanumeric code with control and graphic characters. Table 10.3 shows the ASCII code.

The extended binary coded decimal interchange code (EBCDIC), shown in Table 10.4, is an 8-bit code used with large computers and in particular those produced by IBM and Burroughs. EBCDIC can provide a more extensive character code than ASCII as the 8 bits allow up to 256 characters.

10.8 Error detection and correction

Errors can occur in the transmitted data of every data communication system. Interference factors can be introduced, particularly in the industrial environment, which can cause transmitted 1s to be interpreted as 0s at the receiver, and transmitted 0s to be detected as 1s. The interference factors generally come under the heading of electrical noise, which can be defined as unwanted magnetic or electrostatic fields induced or transferred by some means into the data communicating circuit. Noise generally affects the amplitude of the transmitted data. Electrical noise occurs in two basic forms: natural and manufactured. Natural electric noise is the type of noise generated in nature by phenomena like lightning. This type of noise cannot be controlled, but shielding can be provided. Manufactured noise is the type of electrical noise generated by motors, electrical switching, etc. within the programmable controller environment. Manufactured electrical noise can be suppressed at the source or shielded out of the PLC installation.

Table 10.3 ASCII code

b7					0	0	0	0	1	1	1	1
b6					0	0	1	1	0	0	1	1
b5					0	1	0	1	0	1	0	1
b4	b3	b2	b1	row								
0	0	0	0	0	NUL	DLE	SP	0	@	P	`	p
0	0	0	1	1	SOH	DC1	!	1	A	Q	a	q
0	0	1	0	2	STX	DC2	"	2	B	R	b	r
0	0	1	1	3	ETX	DC3	#	3	C	S	c	s
0	1	0	0	4	EOT	DC4	$	4	D	T	d	t
0	1	0	1	5	ENQ	NAK	%	5	E	U	e	u
0	1	1	0	6	ACK	SYN	&	6	F	V	f	v
0	1	1	1	7	BEL	ETB	'	7	G	W	g	w
1	0	0	0	8	BS	CAN	(	8	H	X	h	x
1	0	0	1	9	HT	EM	)	9	I	Y	i	y
1	0	1	0	10	LF	SUB	*	:	J	Z	j	z
1	0	1	1	11	VT	ESC	+	;	K	[	k	{
1	1	0	0	12	FF	FS	,	<	L	\	l	|
1	1	0	1	13	CR	GS	-	=	M	]	m	}
1	1	1	0	14	SO	RS	.	>	N	^	n	~
1	1	1	1	15	SI	US	/	?	O	_	o	DEL

Table 10.4   EBCDIC code (continued over page).

Decimal	Binary	Hex	Character	Decimal	Binary	Hex	Character
0	00000000	00	nul	41	00101001		
1	00000001			42	00101010		
2	00000010			43	00101011	2B	tab
3	00000011			44	00101100		
4	00000100			45	00101101	2D	carriage return
5	00000101	05	tab	46	00101110		
6	00000110			47	00101111		
7	00000111	07	delete	48	00110000		
8	00001000			49	00110001		
9	00001001			50	00110010		
10	00001010			51	00110011		
11	00001011			52	00110100		
12	00001100			53	00110101		
13	00001101			54	00110110		
14	00001110			55	00110111	37	EOT
15	00001111			56	00111000		
16	00010000			57	00111001		
17	00010001			58	00111010		
18	00010010			59	00111011		
19	00010011			60	00111100		
20	00010100			61	00111101		
21	00010101	15	new line	62	00111110		
22	00010110			63	00111111		
23	00010111			64	01000000	40	space
24	00011000			65	01000001		
25	00011001			66	01000010		
26	00011010			67	01000011		
27	00011011			68	01000100		
28	00011100			69	01000101		
29	00011101			69	01000101		
30	00011110			70	01000110		
31	00011111			71	01000111		
32	00100000			72	01001000		
33	00100001			73	01001001		
34	00100010	22	field separator	74	01001010	4A	£
35	00100011			75	01001011	4B	.
36	00100100			76	01001100	4C	<
37	00100101	25	line feed	77	01001101	4D	(
38	00100110			78	01001110	4E	+
39	00100111			79	01001111	4F	|
40	00101000			80	01010000	50	&

196

Table 10.4   EBCDIC code (continued).

Decimal	Binary	Hex	Character	Decimal	Binary	Hex	Character
81	01010001			126	01111110	7E	"
82	01010010			127	01111111		
83	01010011			128	10000000		
84	01010100			129	10000001	81	a
85	01010101			130	10000010	82	b
86	01010110			131	10000011	83	c
87	01010111			132	10000100	84	d
88	01011000			133	10000101	85	e
89	01011001			134	10000110	86	f
90	01011010	5A	!	135	10000111	87	g
91	01011011	5B	$	136	10001000	88	h
92	01011100	5C	*	137	10001001	89	i
93	01011101	5D	)	138	10001010		
94	01011110	5E	;	139	10001011		
95	01011111	5F	>	140	10001100		
96	01100000	60	-	141	10001101		
97	01100001	61	/	142	10001110		
98	01100010			143	10001111		
99	01100011			144	10010000		
100	01100100			145	10010001	91	j
101	01100101			146	10010010	92	k
102	01100110			147	10010011	93	l
103	01100111			148	10010100	94	m
104	01101000			149	10010101	95	n
105	01101001			150	10010110	96	o
106	01101010	6A	^	151	10010111	97	p
107	01101011	6B	,	152	10011000	98	q
108	01101100	6C	%	153	10011001	99	r
109	01101101	6D	_	154	10011010		
110	01101110	6E	>	155	10011011		
111	01101111	6F	?	156	10011100		
112	01110000			157	10011101		
113	01110001			158	10011110		
114	01110010			159	10011111		
115	01110011			160	10100000		
116	01110100			161	10100001		
117	01110101			162	10100010	A2	s
118	01110110			163	10100011	A3	t
119	01110111			164	10100100	A4	u
120	01111000			165	10100101	A5	v
121	01111001			166	10100110	A6	w
122	01111010			167	10100111	A7	x
123	01111011			168	10101000	A8	y
124	01111100			169	10101001	A9	z
125	01111101	7D	=	170	10101010		

197

Table 10.4   EBCDIC code (continued).

Decimal	Binary	Hex	Character	Decimal	Binary	Hex	Character
171	10101011			214	11010110	D6	O
172	10101100			215	11010111	D7	P
173	10101101			216	11011000	D8	Q
174	10101110			217	11011001	D9	R
175	10101111			218	11011010		
176	10110000			219	11011011		
177	10110001			220	11011100		
178	10110010			221	11011101		
179	10110011			222	11011110		
180	10110100			223	11011111		
181	10110101			224	11100000	E0	blank
182	10110110			225	11100001	E2	S
183	10110111			226	11100010	E3	T
184	10111000			227	11100011	E4	U
185	10111001			228	11100100	E5	V
186	10111010			229	11100101	E6	W
187	10111011			230	11100110	E7	X
188	10111100			231	11100111	E8	Y
189	10111101			232	11101000	E9	Z
190	10111110			233	11101001		
191	10111111			234	11101010		
192	11000000			235	11101011		
193	11000001	C1	A	236	11101100		
194	11000010	C2	B	237	11101101		
195	11000011	C3	C	238	11101110		
196	11000100	C4	D	239	11101111		
197	11000101	C5	E	240	11110000	F0	0
198	11000110	C6	F	241	11110001	F1	1
199	11000111	C7	G	242	11110010	F2	2
200	11001000	C8	H	243	11110011	F3	3
201	11001001	C9	I	244	11110100	F4	4
202	11001010			245	11110101	F5	5
203	11001011			246	11110110	F6	6
204	11001100			247	11110111	F7	7
205	11001101			248	11111000	F8	8
206	11001110			249	11111001	F9	9
207	11001111			250	11111010		
208	11010000			251	11111011		
209	11010001	D1	J	252	11111100		
210	11010010	D2	K	253	11111101		
211	11010011	D3	L	254	11111110		
212	11010100	D4	M	255	11111111		
213	11010101	D5	N				

Note that the EBCDIC code is used for information interchange within the computer system in most applications.

A second cause of errors in data transmission circuits is frequency and phase sensitive components in the data transmission system. The bits of transmitted data are often converted into a frequency where, if the transmitted data is a 0, one frequency is sent and, if the transmitted data is a 1, another is sent. The conversion of the data bit to frequency is made by using a modulator/demodulator (modem) device. A transmission medium such as a land line of physical copper conductor is frequency selective, that is, the transmission medium will have a specific bandwidth (for voice-quality copper conductors, approximately 300 Hz to 3 kHz), and, due to the frequency response of the transmission medium, one of the two frequencies could arrive at the receiver at a different level to the other as the transmission medium may be frequency sensitive. In extreme cases the difference between the level of the two received frequencies may be so great that the receiver cannot compensate for the difference and data will be lost.

A third way that errors can be introduced into the data communication system is by the transmission medium introducing phase distortion onto the data communication circuit. Phase distortion occurs because the propagation time of frequencies is different across a transmission medium which results in the phase relationship of the transmitted frequencies being changed at the receiver. Transmission mediums can be conditioned to reduce frequency and phase distortion by the use of equalizers which cause the transmission medium to have a flat frequency response over the desired bandwidth.

There are many types of error detection schemes that can be used in data communication; two of the more common methods will be examined: parity, and cycle redundancy check.

Parity error detection

The serial ASCII character contains 7 data bits, start and stop bits to frame a character, and often an extra bit, termed a parity bit, which is used to provide some error-detecting ability at the receiver. The data transmitting equipment adds a parity bit to each character. This could be odd parity if the 1s in the character add up to an odd number, or even parity if the 0s in the character add up to produce an even number. The receiver generates its own parity bit based on the received data, and compares the receiver-generated parity bit with the transmitted parity bit. If the bits are the same, the received data is considered valid; if the parity bits are different, the data is rejected as being invalid.

Parity error detection is simple, but errors can be missed if two data bits in the transmitted data are reversed or two 1s are changed to two 0s, as the parity error check will indicate valid data. Figure 10.7(a) shows an ASCII character with parity where a single bit is incorrect, and Figure 10.7(b) shows the same character where two bits are inverted and no error is detected. Note that the start and stop bits are not included in this parity check, but the parity bit is included.

Cyclic redundancy check error detection

The cyclic redundancy check assumes the transmitted data for each character is a single binary number. The binary number is divided by a constant which produces a quotient and a remainder. The quotient is unused and the remainder is transmitted along with the data. At the data receiver, the received data is divided by the same

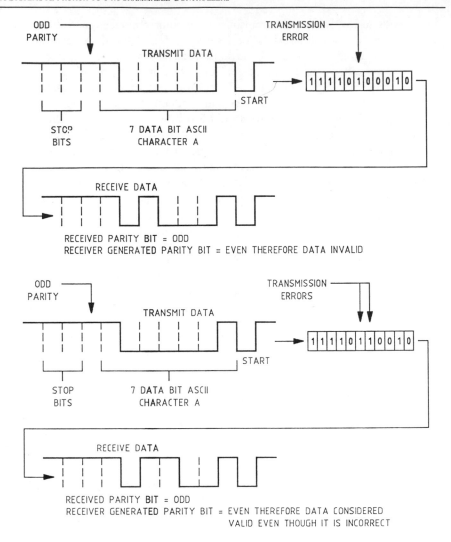

Figure 10.7(a) Parity check error detected   (b) Parity check error undetected

constant and a remainder generated. The received remainder is compared with the remainder transmitted with the data. If the remainders are the same, the data is considered valid. If the remainders are different, the data is considered invalid.

Error correction

Many error detection and correction techniques have been devised for use with data communication, but they all involve the use of redundant data bits which make the transmitted information longer and therefore more likely to contain errors, as well as being slower. The simplest and most common method of error correction is to request the blocks of data to be sent again when errors are detected.

10.9 Modems

The discussion to this point has been largely in reference to baseband data communication where bits of information are placed directly onto the data transmission medium. If data transmission is required between two distant points, or the public telephone network is to be used, the data bits must be converted to audio tones, as data bits represented as voltage levels may be so attenuated as to make the data signal unreadable. A modem is used to convert the transmitted data bits to audio tones. It modulates the signal at the transmitter and demodulates the signal at the receiver to convert the audio tones back to data bits. The term modem is made up of the first two letters (mo) of modulate and the first three letters of demodulate (dem).

Modulation is the superimposing of intelligence onto a carrier. The carrier is the frequency that carries the intelligence from the transmitter to the receiver, and the intelligence is the data bit pattern that contains the transmitted character. The three basic modulation methods used in conjunction with data communication are:

- amplitude modulation (AM) where the intelligence is superimposed onto the amplitude of the carrier, or causes the amplitude of the carrier to change;
- frequency modulation (FM) where the intelligence is superimposed onto the frequency of the carrier, or causes the frequency of the carrier to change;
- phase modulation (PM) where the intelligence is superimposed onto the carrier which causes the phase of the carrier to change.

Figure 10.8 shows the block diagram of a modem using frequency modulation techniques to convert, in one direction, the digital 1s to a 2225 Hz audio tone, and the 0s to a 2025 Hz tone. In the other direction, the transmitted 1s are converted to a 1270 Hz tone, and 0s to a 1070 Hz tone. As two-way communication is necessary, four frequencies are required. Filters are needed within the modem to separate the transmit frequencies and the receive frequencies. The transformer provides impedance matching to the balanced transmission line. Tones for modems are usually within the 300 Hz to 3400 Hz range as this is the operational bandwidth of a telecommunications circuit.

10.10 Data transmission medium

The transmission medium used for data communication can be any electronic circuit used for communications. The channel between the transmitter and receiver may include copper conductors, twisted pairs, co-axial cable, radio links, and optical fibers. The path between the data transmitter and receiver may include privately-owned cabling, public telephone network, and interstate or overseas radio satellite or cable systems.

The vast majority of data communications involving programmable controllers remain within the immediate location of the controlled plant with a central control station or control room. Industry is, however, tending towards providing communications to a central computer to oversee the operation of the entire plant, both technically and commercially, using computer integrated manufacture (CIM), thus sending data communications over greater distances with a wide variety of data communication mediums.

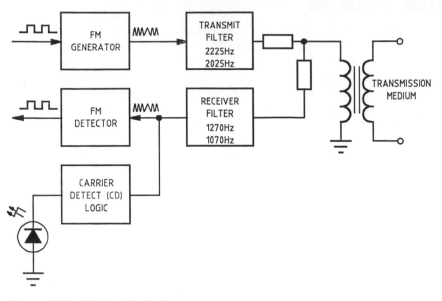

Figure 10.8   Modem block diagram

The simplest means of communication, either serial or parallel, is via copper conductors. The usual means of connecting a parallel printer to a programmable controller, for instance, is to have copper conductors located side by side to form ribbon cable. Approximately 10 meters (30 feet) is the maximum distance that ribbon cable can be used without signal degradation. Signal degradation is largely caused by the shunt capacitance characteristic of the cable distorting the transmitted signal. In addition, ribbon cable is susceptible to picking up unwanted signals (noise) because it is unshielded.

To increase the distance that data can be transmitted, a serial data communication is used. A single pair of wires can convey serial communications over 100 meters (300 feet). The characteristics of the single-pair line can be improved by twisting the pair to provide increased noise immunity. Twisted pairs formed into cables are often used to provide two-wire lines between two locations, but, like side-by-side copper conductors, twisted pairs have a bandwidth limitation of about 3 KHz and therefore have transfer data at rates of approximately 1200 baud without excessive errors.

Co-axial cable is a form of unbalanced data transmission line which can be used to transfer data over relatively large distances, 100 to 150 meters (300 to 500 feet), at fast data rates. This is because co-axial cables have an improved frequency response compared to copper conductors. Because the co-axial cable braid is at earth potential and provides a screen to shield against unwanted signals that may be induced into the cable from the surrounding industrial environment, it is often used to interconnect programmable controllers. Instrumentation cable with a pair of inner conductors and a screen is an alternative to the single inner conductor co-axial cable and is used in applications where the characteristics of a balanced transmission line are required as well as the advantage of cable shielding.

Because of its extremely wide bandwidth, optic fiber cable is used for high capacity data communication with bit rates of up to 140 Mega bits per second. The optic fiber is an excellent replacement for copper conductor systems as it is free from the electromagnetic interference often found in industrial applications. In addition, optic fiber does not radiate signals along its length, or cause interference to adjacent circuits. Increased security is also a bonus as it is difficult to tap into an optical fibercable. It consists of an inner glass core which allows light to pass through, and an outer glass cladding that acts as a guide to the light along the core and minimizes light energy loss. The transmitter in an optic fiber system can be a laser (light amplification by stimulated emission of radiation) device, or a light-emitting diode (LED). The laser or LED is pulsed rapidly to encode the system data. The signal can be detected and decoded at the receiver end of the system by the use of a light-sensitive diode.

10.11 Data communication interfaces

An interface in an electronic system is the point where information is transferred from one point to another. It implies that the data is compatible across the interface, at both the transmit and the receive point. The data terminal equipment (DTE) connects to the data communications equipment (DCE) across the interface at one end of the system, and a DCE to DTE interface at the other end of a two-way system. The data communication system interconnections are under the control of the system designer or manufacturer; interface characteristics are chosen to suit the application. When equipment from different manufacturers is to be interconnected it becomes important to clearly define interface characteristics. The major groups involved in determining the recommendations for interface connections are:

- Electronics Industries Association (EIA);
- Institute of Electrical and Electronic Engineers (IEEE);
- American National Standards Institute (ANSI);
- International Electrotechnical Commission (IEC);
- International Consultative Committee for Telephone and Telegraph (CCITT) European Recommendations;
- International Standards Organization (ISO), a United Nations organization.

The interface connection specifications contain:

- mechanical information such as type of connector, cable connections, etc;
- electrical characteristics such as the type and amplitude of the signal;
- procedural information including the signal timing;
- functional information which describes the purpose of the signals.

RS232c interface

The RS232c interface connection is the most widely used in data communications, including many industrial programmable controller applications. The RS232c connection was originally used to define interface connections between data terminal equipment and modems, but it has also been extensively used as an interface connection for many other applications. As defined by the IEA, it is almost identical to the CCITT V24 interface recommendation.

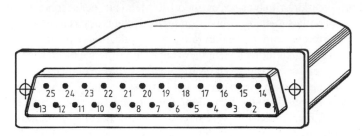

Figure 10.9   25-way D connector

The RS232c connection will be discussed in terms of its mechanical, electrical, functional, and procedural characteristics. Mechanically, the connector used for the RS232c connection is a 25-way D connector as shown in Figure 10.9. The maximum recommended cable length for RS232c is 15 meters (50 feet). Table 10.5 shows the relationship between the connector pin assignment numbers and the function of circuit between the DTE and DCE.

Electrically, the RS232c connection can consist of synchronous or asynchronous transmission with an unspecified code. The RS232c specification refers to signalling rates in baud of 75, 110, 150, 300, 600, 1200, 2400, 4800, and 9600, with logic level in the range of -5 to -15 volts for a logic 0 level, and $+5$ to $+15$ volts for a logic 1 level, with signals between $+3$ and -3 volts being undefined, as shown in Figure 10.10. The capacitance of the cable limits the number of data bits per second that can be transmitted. At 1000 bits per second RS232c signals can be transmitted over cables as long

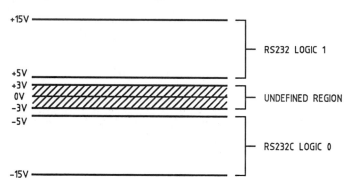

Figure 10.10   RS232c electrical levels

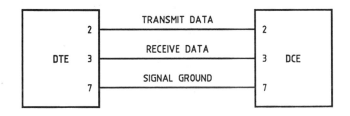

Figure 10.11   RS232c minimum connection

as 1500 meters (5000 feet); at 20 000 bits per second, the cable length is less than 18 meters (60 feet). Table 10.6 shows the electrical specifications for RS232c.

Table 10.5   25-connector pin assignments

PIN No	EIA CCT	CCITT CCT	Signal description	Abbreviations
1	AA	101	Protective ground (chassis)	GND
2	BA	103	Transmitted data	TD
3	BB	104	Received data	RD
4	CA	105	Request to send	RTS
5	CB	106	Clear to send	CTS
6	CC	107	Data set ready	DSR
7	AB	102	Signal ground	SG
8	CF	109	Received line signal detector	DCD
9			Unused	
10			Unused	
11		126	Select transmit frequency	STF
12	SCF	122	Secondary received line signal detector	
13	SCB	121	Secondary clear to send	
14	SBA	118	Secondary transmitted data	
15	DB	114	Transmitter signal element timing	(DCE)
16	SBB	119	Secondary receive data	
17	DD	115	Receiver signal element timing	
18			Unassigned	
19	SCA	120	Secondary request to send	
20	CD	108/2	Data terminal ready	DTR
21	CG	110	Signal quality detector	SQ
22	CE	125	Ring indicator	Ri
23	CH/CI	111/112	Data signal rate selector	DTE/DCE
24	DA	113	Transmit	
25		142	Test indicator	BY

Table 10.6   RS232c electrical specifications

Driver output levels with between a 3 and 7 k ohm load logic 1, +5 to +15 V logic 0, −5 to −15 V

Driver output no load — −25 V to +25 V

Driver output impedance with power off — greater than 300 ohms

Output short circuit current — less than 0.5 amp

Driver slew rate — less than 30 V/ S

Receiver input voltage range — −25 to +25 volts

Receiver output with open circuit input — logical 1

Receiver output with 300 ohm to ground on input — logical 1

Receiver output with +3 volt input — logical 0

Receiver output with −3 volt input — logical 1

Receiver input impedance between 3 and 7 k ohm

Maximum load capacitance — 2500 pF

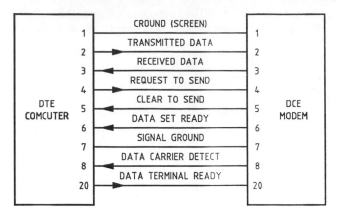

Figure 10.12   Computer-to-modem RS232c connection

The functional characteristics are listed in Table 10.5, but not every circuit need be used to communicate using RS232c. In many applications three wires can be used for RS232c as shown in Figure 10.11. The most often used circuit between a computer as a DTE and a modem DCE is shown in Figure 10.12. The wire labels indicate the function of each circuit and the arrows indicate the signal direction.

The procedural definitions of the RS232c specification provides the timing and control requirements of the signals within the interface. As the requirements will differ between interface connections, a full description of procedural protocol is beyond the scope of this chapter. The manufacturers' handbooks for programmable controller data communication units will provide the necessary information to determine the procedural characteristics of the interface circuit.

Current loop

To connect a teletype to a programmable controller, a current loop can be used. The current loop uses the presence or absence of current to represent 1s and 0s. The current could be typically either 20 or 60 mA, depending on the teletype manufacturer. The current loop can operate at speeds of up to 9600 bits per second, has high noise immunity suitable for the industrial environment, and is used over distances of up to 460 metres (1500 feet).

RS424a, RS423a, and RS449 were introduced, making use of the advantages of the current loop, to overcome the inadequacies of RS232c. RS422a and RS423a cover the electrical specifications of the interface, while RS449 covers the control functions. RS422a allows high transmission rates compared to RS232c because individual balanced wires are used for each signal. As the transmission system is balanced, ground requirements are less critical and the undefined region between a 1 and a 0 is reduced from a range of +3 volts to −3 volts for RS232, to a range of +0.2 volts to −0.2 volts, therefore allowing the $\pm$5 volt power supply commonly used in computers and programmable controllers in data communication circuits.

RS423a uses an unbalanced transmission system with an earth return similar to RS232c. It operates for both RS422a and RS232c circuits.

206

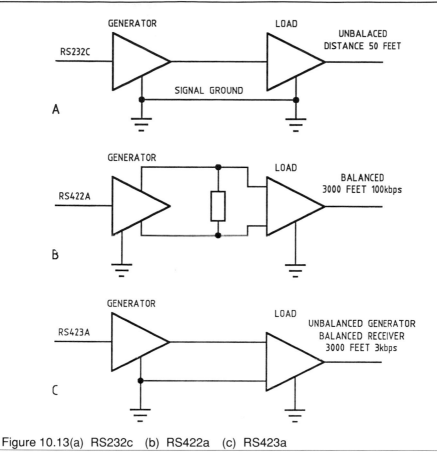

Figure 10.13(a) RS232c (b) RS422a (c) RS423a

RS449 may eventually replace the RS232c circuit because it has the advantage of higher speed and greater working distance, as well as some functional characteristics which allow RS449 to be used in conjunction with automatic modem testing. Figure 10.13 shows the RS232c, RS422a, and RS423a circuit connections.

Centronics interface

The centronics interface is a parallel interface often used to connect printers to programmable controllers. The centronics interface employs a Centronics P/N 31 310012–1016 or Amphenol 57–40 360 or equivalent female connector, as shown in Figure 10.14.

The centronics interface is used to transfer parallel 8 bits of data over distances less than 10 meters (30 feet). A strobe clocks the information from the programmable controller to the printer logic. The logic levels used for the centronics interface are TTL, which are:

- for a high or logic 1, a voltage between +2.4 volts and +5 volts and not exceeding 5.5 volts;
- for a low or a logic 0, a voltage range of 0 volts to +0.4 volts and not exceeding −0.5 volts.

207

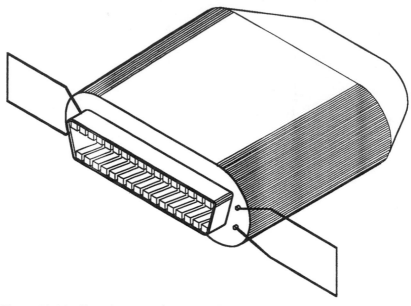

Figure 10.14   Female centronics connector

The current required for the centronics interface is supplying (sourcing) current of 0.320 mA for a logical 1, and inputting (sinking) current of 14 mA for a logical 0. The data transmission rates available across this interface are printer dependent, but speeds of up to 100 000 characters per second are possible.

The functional and procedural characteristics of the centronics interface will differ slightly between systems, in particular according to the type of printer used. The printer manufacturer's handbook will provide details of these characteristics.

10.12 Local area networks (LANs)

Local area networks in a process environment can be defined as communication networks which interconnect programmable controllers and microprocessor-based equipment. The term 'local' means that the plant and equipment is limited to a factory area, and this may mean several acres in larger factories, with an overall distance of no more than 5 kilometers (3 miles). Data transmission rates of 10 Mbps are common, and rates of 100 Mbps are available with undetected error rates of 1 part in 10^{12}. A transmission medium of co-axial cable or optical fibre is used in the industrial environment as it provides high noise immunity.

The ideal local network is defined by the open system interconnection (OSI) model, determined by the International System Organization (ISO), European Computer Manufacturers' Association (ECMA), and CCITT. Each level up the OSI model employs a higher order of supervision. The OSI is a seven-functional layer structure, each layer being a separate unit which provides specialized requirements. Figure 10.15 shows the OSI model. The lower three layers closely represent the physical characteristics of the X25 packet specification which is the interface between a DTE and DCE operating in the packet mode.

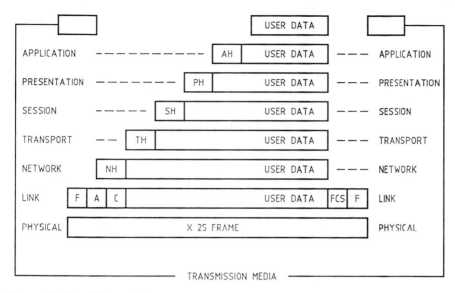

Figure 10.15   OSI model

Packet, as applied to data communication, is a block of data which includes a header or address of the destination of the information. A packet data system is one in which packets of information can be sent from different input sources and not as a continuous stream. As many users can operate in the packet mode simultaneously, system efficiency is increased.

A description of the OSI model layers, from the highest to lowest, is as follows:

- The application layer serves to allow inward and outward communication for a wide range of applications. In a programmable controller, the application layer allows the interconnection of the computer to real time process control programs.
- The presentation layer converts the information from the application to a machine-independent form with character and command format independence.
- The session layer establishes and maintains liaison between the applications, exchanging data by ensuring data reaching the system is routed to the correct application and is synchronized.
- The transport layer, in conjunction with the lower layers, provides a universal transport service independent of the physical medium. Optimum use of the available resources by providing the best quality data transfer is a characteristic of the transport layer.
- The network layer provides for the data transfer between networks via gateways or between network layers.
- The data link layer provides error detection, and possibly correction, as well as frame synchronization across the data medium. In addition, the data layer provides data links between network layers, thus allowing a means to activate, deactivate and maintain data transfer by providing correct procedures and functions.
- The physical layer provides mechanical and electrical connections, allowing the activation and deactivation of the data circuit and fault notification.

209

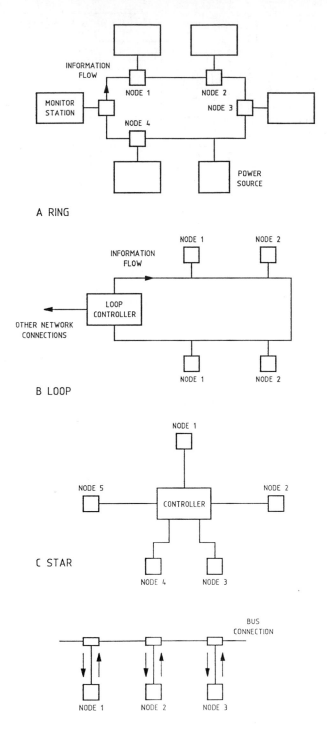

Figure 10.16 Network topologies (a) Ring (b) Loop (c) Star (d) Bus

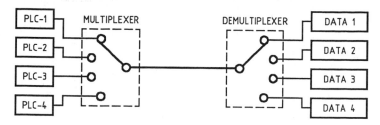

Figure 10.17 Time division multiplexing

The local area network has three descriptive characteristics: network topology, sharing techniques, and signalling methods. The network topology is the physical layout of the interconnected nodes (nodes being the equipment-connecting point to the network). Network topologies as shown in Figure 10.16 are ring, loop, star, and bus.

Network sharing techniques

Time-sharing techniques in data communication systems allow more than one user to be connected to a system and communicate, being transparent to other users, within the system. The most common time-sharing techniques used are:

- time-division multiplexing, and frequency-division multiplexing;
- polling.

Time-division multiplexing transfers the data from each one of a number of data sources to the receiver, one block of information at a time, over a common transmission medium (see Figure 10.17). The multiplexer switches the data to the line and the data must be demultiplexed at the receiver. Time-division multiplexing is used to connect a number of programmable controllers to a single master computer. It is used where low speed programs are in operation as it takes some time to multiplex and demultiplex the information.

Frequency-division multiplexing is used where the transmission medium has a wide bandwidth and the data being transferred from the transmitter to the receiver is using an exclusive frequency bandwidth. A gap between each exclusive band of frequencies is required to reduce the possibility of interference. The advantage of frequency-division multiplexing is that the data path is continuous between the transmitter and the receiver, as shown in Figure 10.18.

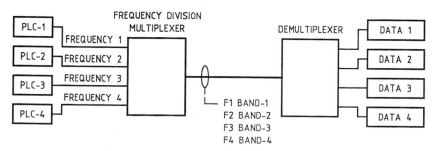

Figure 10.18 Frequency division multiplexing

A polling system is somewhat similar to the time-division multiplexing system, as a polling device is used to call up the information on the remote programmable controller in a cyclic manner. The polled devices each have a unique address. Methods of placing priorities on the polled device allow the polling time to be reduced, or the remote programmable controller to signal the polling device that a change of status has occurred.

Critical process control does not generally lend itself to time-division multiplexing or polling, as real time changes cannot be reflected by time-sharing techniques. The consequences of multiplexer failure must also be considered.

10.13 Distributed-intelligence PLC systems

Programmable controllers are being used in a wide range of industrial and commercial applications, such as batch control, continuous control, and data acquisition. The trend in industry is toward stand alone programmable controllers with data communicating facilities to allow information from the factory floor to be transferred to a central control room where a mini or personal computer is used to supervise the operation. Each programmable controller has sufficient intelligence to continue to carry out the controlled process even if the data link or the host computer fails, therefore making it unnecessary for the host computer to operate in real time. Other advantages of the distributed system is that failure of an individual programmable controller will not necessarily completely stop production because methods of bypassing the faulty area can be designed into the system. A single programmable controller controlling all the processes, however, will stop the process if failure occurs. The cost of the distributed programmable controller system with each unit having less than 128 I/0 is less than an extremely large programmable controller unit, although the cost of an extremely large unit and a large number of smaller units would be comparable. Installation costs, in particular commissioning time, of the smaller units is always an advantage. Cable runs, wiring schedules, and commissioning de-bugging is also much reduced in the smaller system.

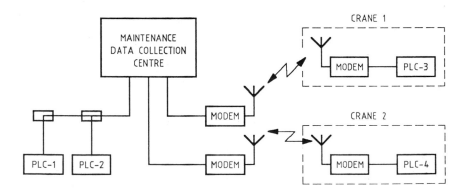

Figure 10.19   Distributed intelligence PLC system

Figure 10.19 shows a distributed-intelligence PLC system incorporating several mobile cranes in a container loading facility. The cranes' PLCs transfer information to a maintenance data collection center where early detection of crane fault warnings reduce loading delays. The system consists of two fixed land line links to programmable controllers involved in automatic loading, and two radio links direct from the crane to the maintenance center. Failure of the individual programmable controller will not cause the overall loading function to cease as the cranes can be manually driven.

10.14 Manufacturing automation protocol (MAP)

Programmable controllers require a data communication facility to communicate between factory floor equipment, such as robot controllers, other programmable controllers, numerically controlled machines, etc., independently of the equipment manufacturer, to allow factory production to become integrated. General Motors (GM) in the USA identified this need. There will be 20 000 systems installed in GM factories by the early 1990s, and as GM are working toward an integrated manufacturing system, a common protocol is required. GM defined a manufacturing automation protocol (MAP), the specification for which was available to anyone, which was based on the layers of the OSI model so as to achieve LAN compatibility. Control and automation system designers and manufacturers are gradually committing themselves to MAP compatibility. Table 10.7 shows the MAP layer standard, which is a 10 Mbps, co-axial medium, token bus system as defined by the IEEE 802–4 standard. OSI layers 1 and 2 are achieved using hardware electronics; layers 3 to 7 are attained by the use of software.

Table 10.7   OSI layer MAP protocol

OSI layer definition	MAP protocol
7 application	ISO file transfer
6 presentation	NUL
5 session	ISO session kernel
4 transport	ISO transport class 4
3 network	ISO internet
	IEEE 802.4 class
2 data link	IEEE token bus
1 physical	IEEE 802.4 broadband

The application layer allows access of user application programs to the system which may be:

- user specific – manufacturing message format standard (GM MMFS);
- application specific – file transfer FTAM;
- common element – common applications service.

213

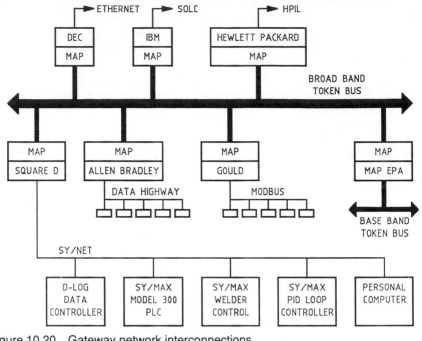

Figure 10.20   Gateway network interconnections

The GM-MAP is providing the foundation for the interconnection of LANs within the manufacturing industry factory environment. Because many LANs have been in operation for some years, using protocols such as Ethernet, and because there is a need to access other proprietary networks, it is necessary to provide a means of interconnection between equipment and networks. The use of MAP gateways performing protocol conversions allows messages to be stored and forwarded from nodes on non OSI networks to nodes on MAP OSI networks. This is achieved with existing equipment (see Figure 10.20).

The square D programmable controller in Figure 10.20 shows a range of equipment that could be networked. Two hundred nodes or more may be connected to a PLC LAN system. Each node must be uniquely identified and requires a network interface module (NIM) to provide the interconnection and high-level data link control (HDLC) to establish data communication, including synchronization, error checking, etc.

The SY/NIM module shown in Figure 10.21 is configured by selecting a unique station number by use of the thumbwheel switches on the module front. The selection of the station number provides a time slot consisting of a single pair of shielded conductors for data transmission over the data communication network. If more than one module was given the same station number, data collisions could result. The unique station number can also be used to prioritize the station, giving a high priority station access to the network before a lower priority station, thereby categorizing the PLCs into those with monitoring functions and those in direct contact with the machine or process control.

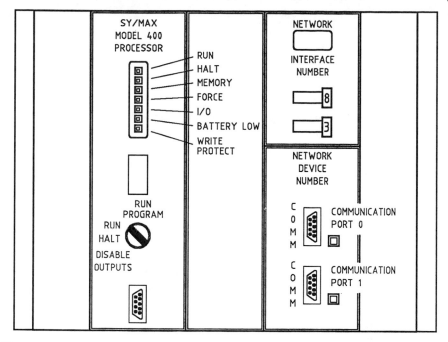

Figure 10.21   Square D 400 series PLC and network interface module

10.15 Test questions

Multiple choice

1 How many data bits make up an ASCII character?

  (a) 5.

  (b) 6.

  (c) 7.

  (d) 8.

2 Which of the following transmission media is considered to have the widest bandwidth?

  (a) Copper conductors.

  (b) Optical fibre.

  (c) Twisted pairs.

  (d) Co-axial cable.

3 The Baudot code for the letter Z has a bit pattern of:

  (a) 00000.

  (b) 10001.

  (c) 01110.

  (d) 11111.

215

4 Which of the following characteristics is *not* found in a two-wire balanced transmission line equivalent?
 (a) Shunt capacitance.
 (b) Shunt resistance.
 (c) Series resistance.
 (d) Shunt inductance.
5 A continuous multiplexed signal between a programmable controller and a master computer can be achieved when:
 (a) time division multiplexing is used.
 (b) polling is used.
 (c) parallel transmission is used.
 (d) frequency division multiplexing is used.

True or false

 6 Digital information is in the form of 1s and 2s.
 7 Parallel data communication is considered to have an advantage over serial communication when a printer is involved, as the parallel communication requires fewer cable conductors.
 8 Baud and bit rate always are the same numeric value.
 9 Full duplex operation allows communication in two directions simultaneously.
10 Asynchronous transmissions require the clock to be transmitted along with the data signal.

Written response

11 Describe the operation and function of a modem.
12 Describe the function of the parity bit in a serial data transmission with respect to error detection.
13 Sketch an RS232c interface that could be used to connect a DTE to DCE using minimum connecting wires.
14 Discuss the advantages and disadvantages of RS232c compared to the current loop.
15 Describe the advantages of using a local area network (LAN) in the process environment.

CHAPTER ELEVEN

Computer-controlled machines and processes

11.1 Computers in process and machine control

Flexible manufacturing systems (FMS) are evolving as microprocessor-based electronic equipment and machines used in manufacturing are developed with data communication functions and compatibility which allow the individual devices to be integrated into a single system. Initially, data communication systems were provided between programmable controllers, but it was apparent that the advantages of integrated manufacture required computer-controlled and numerically-controlled machines and robot controllers to be connected to the programmable controller, which in turn would interconnect with a host computer and other equipment via the factory local area network.

A flexible manufacturing system area, containing machines, is termed a work cell. A work cell with associated machines is shown in Figure 11.1. The main difference between the programmable controller and the cell controller is the language which is used to program the cell controller. The PLC programming language is simple and requires limited programming knowledge; the cell controller requires the programmer to have more programming knowledge. Many software packages are available for cell controllers which will enable the cell to integrate with the plant's existing data communications system.

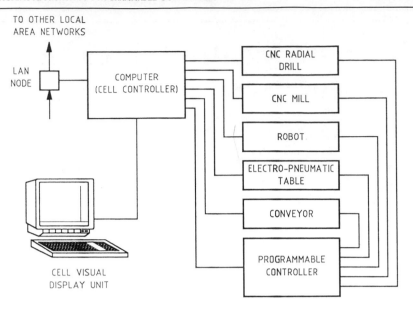

Figure 11.1   Flexible manufacturing system cell

A host computer is used to fully integrate a number of cells into a computer integrated manufacturing (CIM) plant. Usually the computer used is the mainframe or minicomputer used for administrative purposes in the factory. The integrating of plant floor and administrative procedures makes substantial improvements to plant productivity by allowing:

- the implementation of just-in-time (JIT) stock control;
- automatic billing, despatch, and inventory control;
- flexibility of products manufactured;
- on line plant-wide communication;
- distributed hierarchical control of the manufacturing plant.

Figure 11.2 shows the block diagram of a CIM system where four hierarchical levels of control are used. The number of levels of control used is largely dependent on the computing power and the demand on controller availability at any given level.

The use of personal computers as controllers and program terminals for programmable controllers is common. Software available from programmable controller manufacturers is extensive and extremely helpful during program design. The majority of software available is designed to run on IBM* or compatible personal computers.

11.2 Fundamentals of computer operation

Because personal computers, sometimes termed microcomputers, are being used extensively in industry as program terminals or to directly interface to the industrial environment, an overview of IBM or IBM-compatible personal computers, and of the

*IBM is the registered trade mark for International Business Machines.

218

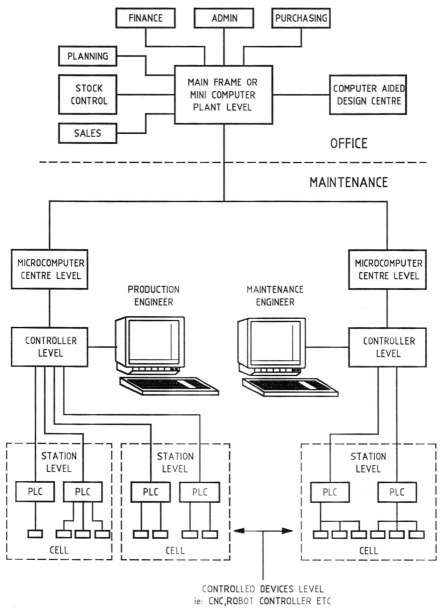

Figure 11.2   CIM system

problems which may be encountered when interfacing the computer to the process or machine, has been included.

All computers consist of two basic components: hardware, and software. The computer hardware is the physical component which makes up the computer system and includes:

- monochrome or color monitor;
- processor and memory;

- keyboard;
- disk drives;
- peripheral devices, such as printers, plotters, etc.

The software contains the program information which makes the computer operate. It consists of:

- operating systems;
- system utilities;
- applications programs.

The programs are usually stored on some form of mass storage system, such as magnetic tape or floppy disk, and loaded into the computer's random access memory (RAM) as required.

The block diagram of a computer system is shown in Figure 11.3 where the central processing unit, which is microprocessor based, is used to control the other components within the system using data communication techniques. For a computer to be useful, information must be input by the operator, the results output by the computer, and the information stored on some form of retrievable mass storage system. The size of the RAM necessary to operate the personal computer as a PLC program terminal may be as much as 640 k, which is 640 kilobytes of storage capacity. As there are 1024 bytes to one kilo, 640 k is equal to 655 360 bytes of storage space. The computer can be interconnected to other computers via a local area network to maximize its use.

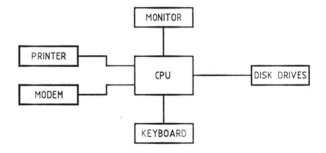

Figure 11.3   Basic computer system

Mass storage systems

The mass storage systems used with personal computers may be:

- $5^1/_4$-inch floppy disk;
- $3^1/_2$-inch floppy disk;
- hard disk.

Combinations of disk types can be used to make up the computer system. Although some disk systems require 8-inch disks, they are not used extensively.

The diskette, usually termed disk, is a magnetic storage device on which information is stored for later retrieval. The information is stored on tracks in sectors of the disk; the location of the information is configured to allow the computer access to a sector to read the information. The disk must be formatted into tracks and sectors using the disk operating system FORMAT command prior to it being used for program storage. Figure 11.4 shows a $5^{1}/_{4}$ inch disk with the arrangement of tracks and sectors.

The disk is inserted in the disk drive with the slot and labels up. The disk drive door must then be shut to allow the disk reading heads to read the information. Care must be taken when handling the disk or the information stored on the disk may be corrupted:

- Never touch the exposed disk at the head slot, or surface.
- Never bend the disk.
- Never expose the disk to heat or excessive sunlight.
- Never expose the disk to electromagnetic or electrostatic fields that are likely to occur in the industrial environment.
- Use only soft tip pens when labelling the disk.
- Never force the disk into the disk drive.

When not in use, the disk should always be stored in a protective envelope away from electrostatic, electromagnetic or magnetic fields. Always make backup disks, i.e. copies, and store the original disks safely.

Personal computers with a hard disk mass storage system can store 40 million (mega) bytes of information. The hard disk is permanently located within the computer and can store a large number of programs simultaneously. The contents of any floppy disk can be relocated onto the hard disk, and the programs then used as required, keeping the floppy disk as backup copies. The disk drives are identified by a letter, A, B, C, or D. In most systems A and B are the floppy disk drives, and C is the hard disk.

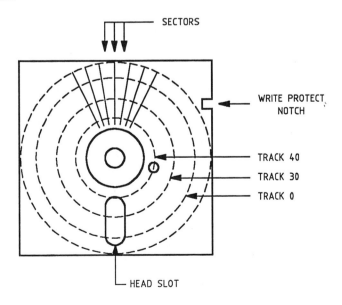

Figure 11.4 $5^{1}/_{4}$ inch floppy disc.

Disk operating system (DOS)

The disk operating system is usually provided by the personal computer vendor. It is used to co-ordinate the inner working of the machine by handling internal housekeeping, retrieving records from the tape or disk, and communicating with the world outside the computer, such as visual display, printer, etc. The DOS allows the operator to:

- read, write and edit files;
- backup files;
- organize a disk;
- manage files;
- display the contents of a disk;
- view error messages.

A file is a collection of stored data. When information is stored on the disk the file is saved or written to the disk. When the information is being used, the disk is being read or data is being loaded into the computer's memory.

Programs are a set of computer instructions which control the computer's operation. They can be loaded into the computer memory from a mass storage system, or saved from the computer memory to a mass storage system. When a program is loaded into the computer memory, the program makes the computer behave in a particular fashion. When turned on, the computer will attempt to determine if a DOS disk containing the system files is in drive A, the default drive. If there is no disk in drive A, the computer will look for the system file on the C drive, if a hard disk is fitted. Failure to find the system file on either drive will result in an error message being displayed on the VDU. The system file allows DOS to be loaded into the computer's memory. If a disk in A drive does not have DOS the computer will display an error message.

The process of starting the computer is often termed 'booting' the system. When the computer is booted up (turned on) with a DOS disk in drive A, a date followed by a time message will be displayed on the screen. The date and time can be entered to set the computer's internal clock or can be bypassed by pressing the RETURN, or ENTER (↵) button. When the prompt A> is displayed, DOS has been loaded into the memory and the computer is now ready to receive your commands. Unless the software provided for use with the programmable controller is self booting, the DOS system will have to be loaded prior to running the PLC software program. To run the PLC software, the software file name is entered via the keyboard and the ENTER button pressed. For further information on the operation of DOS consult the DOS handbooks provided with the computer.

11.3 Computers in the process environment

The personal computer is a complex and delicate electronic device which is sensitive to the environment in which it is placed. Unlike the programmable controller, the computer has not been designed for an industrial environment. Placing the computer in an unsuitable environment may result in computer failure due to hardware or software contamination. In most applications where a computer is either used in conjunction with a programmable controller or is the controlling device, it is desirable to locate the computer in a clean area and communicate with the other devices in the system using data communication techniques.

The types of contaminants that occur can be manufactured or natural. They can be classified as:

- electromagnetic interference;
- electrostatic interference;
- airborne particles;
- temperature extremes;
- chemical corrosive, liquids, and gases.

If the computer is located in a safe environment and data communication techniques are used to communicate with other devices in the system, the most likely entry point for contaminants is via the power supply to the computer and the data communication cables interconnecting the system. The incoming power can be provided with filtering devices to reduce the effect of power supply fluctuations, and, provided the data communication cable is correctly shielded, contamination will be kept to a minimum.

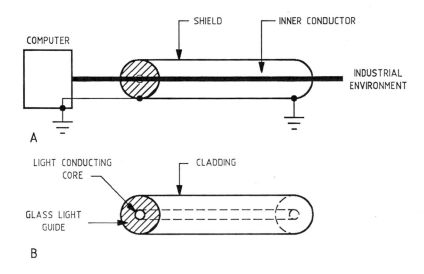

Figure 11.5(a) Co-axial cable   (b) Fibre optic cable

The use of co-axial cable for data communication is preferred due to the relatively low cost of the cable, the high data transfer rates achievable, and the shield that can be provided with the cable. Figure 11.5(a) shows a co-axial cable with an overall shield.

An alternative to copper pair cable is a fiber optic cable where light is used to transfer information along the core of the cable (see Figure 11.5(b)). The advantages of fiber optic cables in industry are:

- they have high electromagnetic and crosstalk immunity;
- there are no ground loops;
- they are safe in hazardous areas, such as explosive atmospheres;
- they have a large bandwidth;
- they are small in size, lightweight, strong and flexible;
- they suffer minimally from temperature variations.

11.4 Process and machine control computer application

The first application discussed is where a computer is used as part of a data logging process in a mass flow gas metering station. A block diagram of this system is shown in Figure 11.6, where analog and digital information is fed into the computer which carries out complex calculations and stores data for a period of time. In this application the computer is housed in a laboratory which is free from contaminants. The analog and digital inputs are fed directly into the computer via analog and digital input cards fitted to the computer. Computers provided with system control and data acquisition (SCADA) software and compatible PLC communication protocols can be used as powerful information collection and processing units.

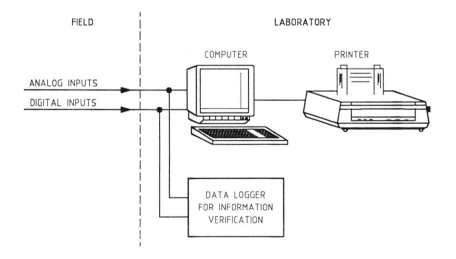

Figure 11.6   Computer data logging

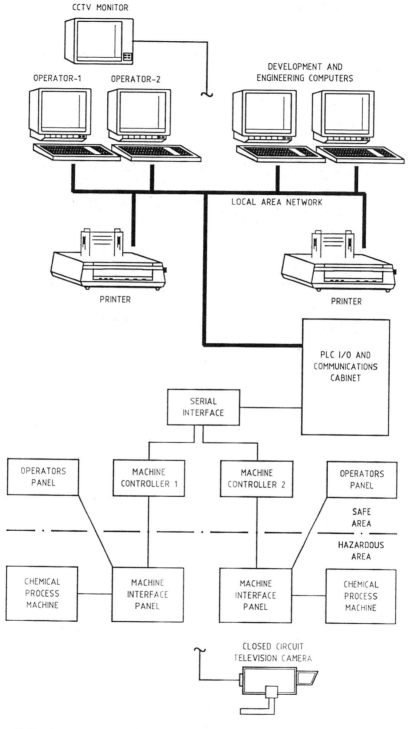

Figure 11.7   Computer-controlled chemical process

The second application is shown in Figure 11.7 where the computer is used to control the operation of a chemical manufacturing process. The computer in this case may be a minicomputer; it uses a local area network to advise control personnel and others of the status of the operation. The advisor system is provided to the operators as mimic screen graphics on the operation console. A serial interface connects the computer to the machine controllers, which may be PLC, electromechanical, or electrical controllers. The machines involved in the chemical making process are in a sealed environment. While the chemicals are being manufactured, the operator is outside the room viewing the process through a television network.

The third application is where the computer is used with a remote program terminal and monitor for a networked PLC system. The block diagram for the system is shown in Figure 11.8. A data bus is provided between the programmable controllers, communications module, and the computer. Vendor software allows the PLC programs to be designed and documented, and provides a monitoring point for the system. When the system is operating correctly, the computer can be used for other applications.

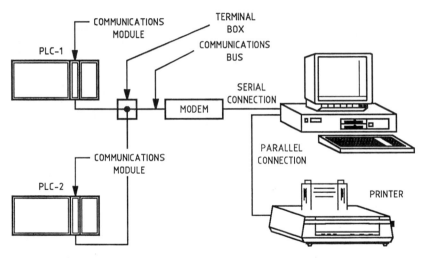

Figure 11.8   Computer as a remote PLC terminal

11.5 Software for PLC program design

Most programmable controller vendors will provide software compatible with a personal computer which will assist in program design documentation and diagnostics. Many of the packages are compatible with available CAD packages. This allows full computer-aided design for the PLC system to be used. The software allows the computer to emulate the PLC program terminal so that program design can take place on the computer, then it downloads the program into the PLC memory via a data communication network. Likewise, a PLC program can be unloaded from the PLC to the computer where program editing takes place. The software provided is making multiple copies. The software is menu driven, which means that the required function is selected from a list displayed on a screen. The cost of the software will vary from several hundred to several thousand dollars.

226

11.6 Test questions

Multiple choice

1 Computer systems consist of two basic components which are:
   - (a) hardware.
   - (b) software.
   - (c) keyboard.
   - (d) disk drives.
2 The information on floppy disks is stored in:
   - (a) heads and slots.
   - (b) sectors and slots.
   - (c) slots and tracks.
   - (d) tracks and sectors.
3 The information stored on a floppy disk can be corrupted by:
   - (a) touching the disk surface.
   - (b) exposing the disk to magnetic fields.
   - (c) operating the disk drive from an AC supply.
   - (d) storing the disk in a protective jacket.

True or false

4 Optical fiber has higher noise immunity than co-axial cable.
5 A flexible manufacturing system cell may contain a number of machines, such as robots, CNC and programmable controlled conveyors.
6 Computer integrated manufacture (CIM) systems do not interconnect the factory floor system to the administrative system.
7 Software is the program information which causes the computer to behave in a particular manner.

Written response

8 Briefly describe the function of the DOS with respect to a computer.
9 State the precautions required when a computer is to be used for an industrial application, and where the computer is located in a clean environment.
10 Describe the advantages that may be achieved by using a personal computer to emulate a PLC program terminal.

CHAPTER TWELVE

Computer numerical control

12.1 Programmable control/numerical control

The contents of this chapter are intended to provide the reader with an introduction to the fundamentals of numerical control with an emphasis on computer numerical control applications. This chapter describes the link between automated processes other than those that exclusively contain programmable controllers. The recent introduction of computer numerical control (CNC), as well as programmable controllers being used for multiple robot system supervision, has brought CNC and PLC devices even closer together; many applications now incorporate both CNC and PLC in an integrated system.

Numerical control devices are driven over a continuous path or to various points to manufacture a component or device using a user program which numerically co-ordinates the machine movement and operation. Numerical control machines are program-dependent and bear a close relation to programmable controllers.

Numerical control has been available to industry since 1949 when it was designed and developed for the aircraft industry. It was probably largely responsible for the automation of industry after 1949 which resulted in the mass production of components and the use of large scale assembly lines. In the early 1980s numerical control suffered a temporary decline in popularity as manufacturing techniques were of the type that did not require production lines, but it regained popularity in the late 1980s and is now considered to be a new and more flexible manufacturing tool within computer integrated manufacturing systems.

12.2 Introduction to numerical control

Numerical control is defined by the Electronics Industries Association (EIA) as:

A system in which actions are controlled by the direct insertion of numerical data at some point. The system must automatically interpret at least some portion of this data.

In general terms, this EIA definition defines numerical control as a flexible method of automatically controlling machines with the use of numerical values; the numerical program controls the machine movements. Originally the program was stored on punched paper tape, but in recent times it has been stored using some form of mass storage such as magnetic tape, floppy disk or solid state memory (e.g. RAM or ROM).

A set of commands make up the NC program and direct the machine to orientate a cutting tool with respect to a workpiece, select different tools, control cutting speed, and direct spindle rotation, as well as perform a range of auxiliary functions such as turning on or off coolant flow.

Numerical control, like programmable control, provides an automated and flexible approach to manufacturing, therefore allowing more scope for advanced planning. The advantages of numerical control over manual control are:

- it provides automation for manufacturing;
- it has a stored re-usable program;
- it is faster and more consistently accurate;
- it is flexible with respect to product;
- it is economically suitable for short or medium runs of product or even manufacturing single complex units.

12.3 Fundamental concepts of numerical control

Figure 12.1 shows a block diagram of a numerical control system. The numerically controlled machine has a system controller which provides signals to either electrical or hydraulic motors which cause the machine head to be driven left or right, as well as up or down. The table can be motor driven to move the workpiece back and forth. The system controller is usually a microprocessor-based system which electronically stores and reads the program and converts the program information into signals which drive motors to control the machine tool.

Feedback, which is defined as some part of the output signal being fed back to the input, is provided to the system controller so that the exact location and speed of the machine head, and therefore the tool, is known, so that it can be exactly positioned to carry out the required function. As the accuracy of the electronics exceeds the mechanical accuracy of the system, the repeatability, and therefore consistency of the manufactured components are within very tight tolerances.

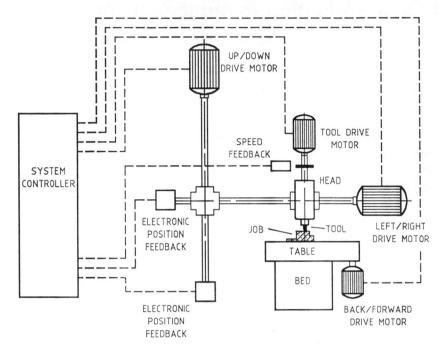

Figure 12.1   Components of a numerical control system

The control system is made up of solid state modules. The tape reader, although still in use, has been largely replaced by other forms of mass storage because paper tape reading techniques are relatively slow (500 characters per second using optimum operating techniques). If it is assumed that the controlled machine is a three-axis type, the location address of the tool is prefixed with the letter X, Y, or Z. Using X, Y, and Z co-ordinates, the machine can be directed to the correct location.

The workpiece is located by using Cartesian co-ordinates, sometimes termed rectangular co-ordinates, which are two axes over which a grid is located. The vertical axis is the Y axis and the horizontal axis is the X axis. The point where the two axes cross is the zero or origin point. To the left of the point of origin on the X axis and below the point of origin on the Y axis, locations are written preceded by a minus (–) symbol. Above the point of origin on the Y axis and to the right of the point of origin on the X axis, locations are written preceded with a plus (+) symbol.

Figure 12.2 shows how points can be located using Cartesian co-ordinates. Point A is located using the co-ordinate system at point X = +2 and Y = +4. Points B, C, and D are located as:

B is X = –3 and Y = +1
C is X = –5 and Y = +4
D is X = +3 and Y = –3

The Z axis of motion is parallel to the machine spindle and defines the distance between the workpiece and the machine.

230

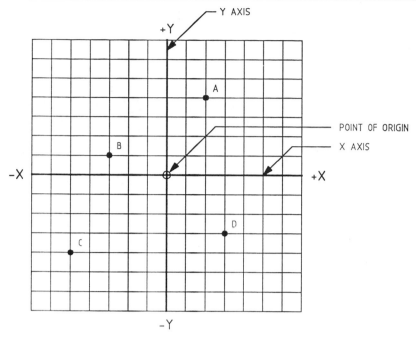

Figure 12.2   Cartesian co-ordinates

12.4 Types of NC programming

The programming of NC machines can be categorized into two main areas: point-to-point programming, and contour programming. An example of point-to-point programming is where holes are to be made, i.e. drilled, punched, bored, reemed, or tapped. The point where each hole is to be located is identified using X and Y co-ordinates. After each hole is drilled the machine is instructed to move to the next point where another hole is to be drilled, etc. The holes are drilled sequentially until the drill program is completed. The path that the machine takes between holes is not important as the tool is in the air between hole locations. The depth of each operation is controlled by the Z axis.

Contour programming, sometimes termed continuous path programming, is used where the workpiece and the machine tool are in continuous contact. The path the tool follows is defined by the program and must be accurate. Contour programming to drill a hole at point A using an NC program may be written:

No 1   X+2000Y+4000F400M3
No 2   12M07F1000

No 1 moves the tool to the correct location followed by turning on the spindle and coolant. No 2 is the next step in the program, preparing the tool for the next function.

The method used in contour programming to obtain continuous machining is based on a process termed interpolation. NC machines may have three types of interpolation: linear, circular, and parabolic. Linear interpolation is where the machine tool is

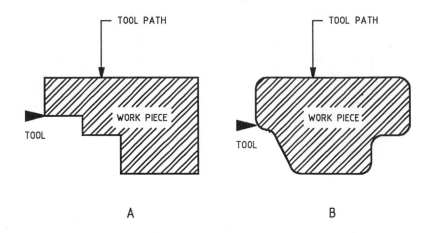

Figure 12.3   Contour programming   (a) Linear interpolation   (b) Linear and circular interpolation

in continuous contact with the workpiece and moves in straight lines between one point and the next. Figure 12.3(a) shows a workpiece which was cut using linear interpolation.

The programming of arcs and circles requires circular interpolation and produces a workpiece as shown in Figure 12.3(b). Circular interpolation requires considerable time to perform the calculations required to move the tool in circles or arcs because each incremental movement must be defined.

Parabolic interpolation is a complex operation, similar to circular interpolation, that can be programmed into some NC controlled machines.

There are three methods used to develop programs for NC machines: manual, computer-aided programming, and NC machine supplier or program writers. The manual method requires the programmer to be able to interpret blueprints or drawings of the part to be machined. The programmer then converts the location of actions to co-ordinates and includes all auxiliary functions. The program, once written, is then typed into a program terminal to produce a punched tape, or is recorded onto some form of mass storage. The program is then verified and, when correct, placed into operation to control the machine's movements.

As many manufacturing industries are using in-house computers it is logical to use the computer as a programming tool for numerical control as well as programmable control applications. The program design can be carried out on a personal computer or a mini/mainframe computer on a time-sharing basis. A numerical control programming language would have to be used to design the program. This simplifies the program writing as the computer will calculate the machine co-ordinates. Using a computer to assist in program writing allows the program to be stored on a convenient mass storage system, retrieved and edited as required, tested off line prior to using the program to control a machine, and plotted using a plotter connected to the computer to assist in program de-bugging.

The use of an NC program language such as APT (automatically programmed tools) allows the NC machine to be programmed using recognizably English words to define the machine's operation. An example of an APT program is shown in Figure 12.4, where the NC machine is directed to point A, turns right along line HL1 and rounds the corner between HL1 and VL1, taking the tangent path between these points.

The programming language handles geometry commands such as POINT, LINE, CIRCLE, and PLANE as well as motion statements such as GO TO, GO ON, GO PAST, and GO DLTA. Instructions which produce the same action in an NC system include FEDRAT AUXFUN DELAY etc.

NC machine vendors, or NC machine programmers, can design programs for NC machines, to the user's specifications. Using outside personnel to design programs will usually result in an efficient, ready-to-use program in a short time. If the program is designed by those maintaining the NC system, however, there will be a better understanding of the machine's operation and therefore reduced down time due to faults.

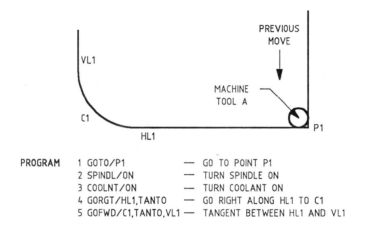

Figure 12.4   Computerized numerical control block diagram

12.5 Computerized numerical control

Computerized numerical control was introduced as an option to the hardwired machine control unit. CNC introduced a new flexibility into the manufacturing industry as the microprocessor based control equipment brought new features to numerical control machines, such as:

- improved program mass storage, disk instead of punched tape;
- ease of editing programs;
- axis calibration, allowing repeatable machine motion accuracy;
- reusable machine pattern, which could be stored and retrieved as required;
- information management, where shop floor information such as machine usage can be used to assist in production planning.

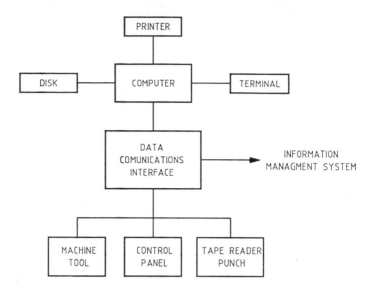

Figure 12.5   Open-loop NC system

Figure 12.5 shows a block diagram of a computerized numerical control system, where the computer is the focal point of the system and controls all aspects of the machining process. The computer could also be used to control other numerical control machines, programmable controllers, and other sections of the manufacturing plant as part of a computer integrated manufacturing system.

12.6 Servo control

Servo mechanisms and control are a major part of CNC and also robotics. They are mechanical, electrical, pneumatic, or hydraulic components used to control the position of a movable machine. Servo control has two basic types: open-loop control, and closed-loop control. The open-loop control system has no feedback to ensure that the actual result of the operation is the same as the expected result. In general, open-loop control is simple and less costly than closed-loop control.

Figure 12.6 shows the basic block diagram of an open-loop control system where the machine table position is assumed to have arrived at the correct location after the command signal has been applied for a specific duration. The location of the table is determined by the number of motor steps of equal angular distance required to move the lead screw to the new location. The motors used in open-loop control are of the stepping variety.

Closed-loop control systems use encoders or resolves to provide a feedback signal which is used to exactly locate the machine position by comparing the fed back signal with the command signal. Figure 12.7 shows the block diagram of a closed-loop NC system where both position and velocity of the table are controlled.

234

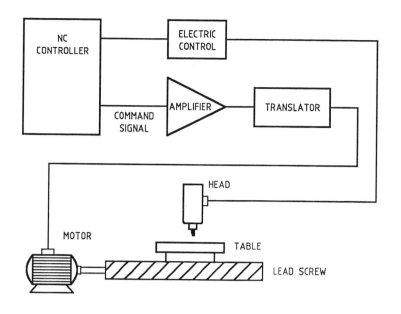

Figure 12.6   Closed-loop NC system

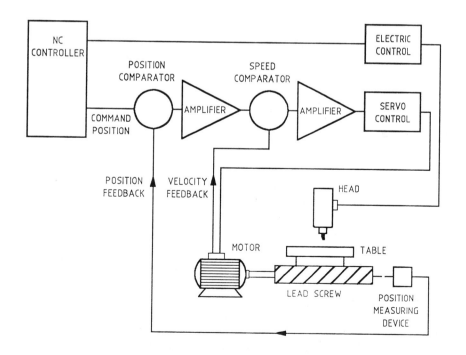

Figure 12.7   Programmable control/ numerical control interface

235

Rotary feedback devices

Rotary feedback devices can be either digital or analog. Digital feedback devices are currently preferred, however, due to the compatibility of the feedback signal to the overall system, its low cost, and its excellent accuracy. One form of digital encoder is the optical encoder, which consists of a light source and rotating disk – a common form of encoder used for servo control.

Resolves are mechanical-to-electrical transducers which produce an analog output with a value dependent on the input signal derived from the lead screw. Many resolvers are rotary transformers where the coupling between the windings is varied by rotating one set of windings with respect to a stationary set of windings.

12.7 Programmable control/numerical control application

Figure 12.8 shows a block diagram of a single-axis numerical control machine which receives input signals from a programmable controller. The lead screw is driven by a three-phase AC motor. The speed of the motor is controlled by predetermined signals generated within the NC system. The programmable controller interfaces between the NC system and other controlled devices.

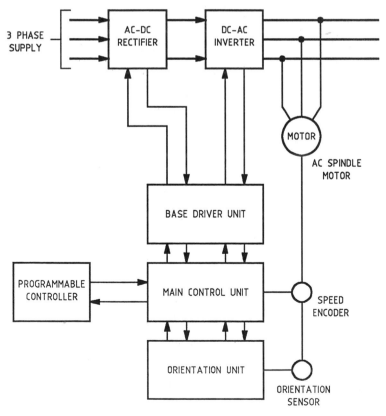

Figure 12.8   Cartesian co-ordinate robot

12.8 Robotics

Robots used by industry are a type of automated machine which, when used with other robots in a fully-automated production system, can be initiated and sequenced by programmable controller. Some difficulty arises when attempting to define a robot; it is generally defined as an automated machine which will respond to input received from the environment in which it is located. Robots most often used in industry are of the type that simulate the functions and structure of the upper human body, i.e. the waist, shoulder, elbow, wrist, and gripper, which are the components of the robot arm.

The Computer Aided Manufacturers International (CAM–I) in the USA defines the aspects of an industrial robot arm as 'a device that performs functions ordinarily ascribed to human beings, or operates with what appears to be almost human intelligence'. A further definition aligns the robot to an automatic machine by stating that a robot is 'an automatic machine with a certain degree of autonomy, designed for active interaction into the environment'.

Robots are classified on the basis of:

- structural configuration and motion;
- trajectories;

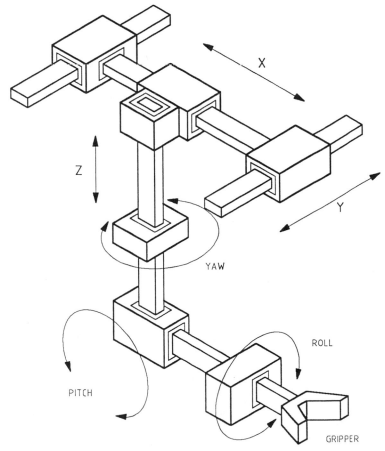

Figure 12.9   Cartesian robot

- performance characteristics.

The robot motion is defined in terms of three basic movements:

- swivel, which is rotation about the longitudinal axis between two joints;
- bend, which is the rotation about the transverse axis of a joint;
- linear, which is motion along a longitudinal axis an extension or contraction.

From robot joint configurations, the five types of robot can be defined:

- revolute or jointed arm type;
- polar or spherical;
- Cartesian or rectangular;
- cylindrical;
- selective compliance assembly robot arm (SCARA).

Figure 12.9 shows the configuration of the Cartesian or rectangular robot, with three translatory motions X, Y and Z directions, which could be controlled by a numerical or programmable controlled system. The revolute robot arm shown in Figure 12.10 best simulates the human arm. This six-axis robot is often controlled using a robot controller, but sequenced using a programmable controller.

Classification of robots based on path control

There are two types of path control: point-to-point, and continuous path. Point-to-point programming causes the robot to shift from a programmed point to the next programmed point, with a pause between each movement. Continuous path robot programming causes the robot to move in a smooth continuous manner along a defined trajectory. The continuous path programmed robot calculates the path using interpolation, which, compared to point-to-point programming, may require 15 per cent to 25 per cent speed reduction because of the time the controller takes to calculate the robot trajectory.

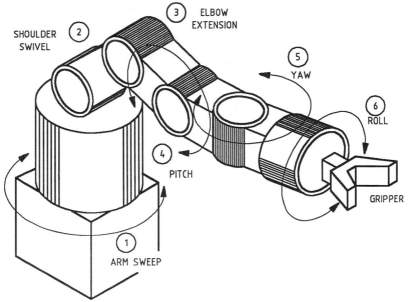

Figure 12.10   Six-axis revolute robot arm

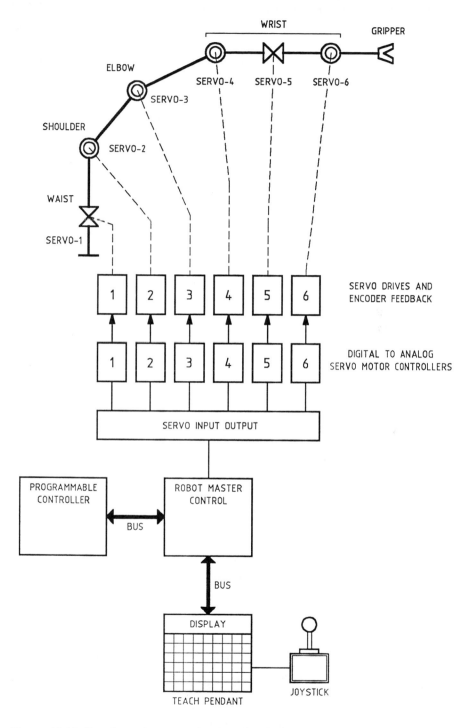

Figure 12.11 Revolute robot control system block diagram

Classification of robots based on performance

Robots can be classified in terms of accuracy and repeatability. Accuracy is the ability of the robot to move to the command position at a specific velocity within a specified tolerance. The accuracy of the robot is usually measured at the end of the arm. Repeatability is the ability of the robot to position the manipulator at a fixed location a number of times. Repeatability is the measure of the location that the manipulator arrives at after a number of attempts.

The block diagram of a programmable controlled robot system is shown in Figure 12.11. In this system, the programmable controller is used to interface with the robot master controller, providing signals to start and stop, and generally controlling the sequence of operation. The programmable controller could be used to directly control the robot servo motors.

The robot program is held in RAM or ROM in the master controller; once the programmable controller signals the robot to commence a sequence, the robot program will control the servo motor's speed and position. The servo motors are provided with encoded feedback in a closed-loop configuration to assist in providing accurate positioning of the gripper and high sequence repeatability.

The teach pendant is used to teach the robot the motion steps that are required to complete the function.

12.9 Test questions

Multiple choice

1 Feedback is used in NC systems to:
   (a) improve system stability.
   (b) ensure the expected machine position is the same as the actual position.
   (c) reduce the system cost.
   (d) improve the system's mechanical accuracy by adjusting the head position.

2 The types of programming used in NC systems are:
   (a) contour.
   (b) continuous.
   (c) point to point.
   (d) ladder logic.

3 The types of interpolation used when contour programming an NC machine are:
   (a) linear.
   (b) non linear.
   (c) arcular.
   (d) parabolic.

4 Servo control may be:
   (a) open loop.
   (b) delta loop.
   (c) token ring.
   (d) closed loop.

5 The teach pendant used in conjunction with a robotic system is used to program the:

(a) programmable controller.

(b) numerical controller.

(c) servo drives.

(d) master control unit.

True or false

6 NC machines are program controlled.

7 The electronic accuracy of a numerical control system is better than the mechanical accuracy.

8 CNC systems are microprocessor based.

9 A numerical control machine can be controlled by a programmable controller, but a robot must be controlled by a numerical controller.

10 Revolute robots are always six-axis devices.

Written response

11 Discuss the features of CNC that are not always available on conventional NC systems.

12 Describe the advantages and disadvantages of open-loop as compared to closed-loop servo systems.

13 Discuss the differences between resolvers and encoders.

14 List four types of robot-based or structural configuration.

15 Write a statement that defines a robot.

CHAPTER THIRTEEN

Programmable controller installations

13.1 PLC environment considerations

Deciding whether or not to automate or control a process or machine by using a programmable controller requires consideration of two main areas:

- PLC needs analysis, consisting of technical and economic considerations;
- PLC integration into the workplace, consisting of human factors and training requirements.

The technical requirements analysis can be carried out by in-house technical experts, programmable controller suppliers, or independent technical consultants. An overview of the required system, without being equipment-specific, is useful and is available as a service from independent technical consultants. During the technical section of the needs analysis, consideration must be given to a range of alternative approaches to automation control system design, as well as programmable control.

The programmable controller system must be able to perform the required functions, and therefore the functions must be able to be clearly defined. The physical location of the PLC must be appropriate with respect to the controlled machine or process. The PLC system must be considered in terms of future requirements. The economic needs analysis should consider PLC system development, installation, and operating costs against increased productivity, benefits from the system, and resultant income.

Human factors must be considered in terms of the impact that automating a process or machine will have on machine or process operators and technical support staff.

Training requirements must be considered, as the installation of 'high tech' equipment will require at least some initial training for plant operators and technical support staff. In fact, the high reliability of programmable controllers means that technical support staff require ongoing training; so few faults occur in the PLC equipment that refresher courses are required to keep the staff up-to-date with the equipment.

13.2 Programmable controller system facilities

The selection of the programmable controller manufacturer who can best supply system requirements is difficult for the first-time user. Most programmable controller manufacturers will supply a device which will have the desired technical features, but it may be necessary to ensure that the equipment supplier can also provide the following services:

- system design assistance;
- full system documentation, including handbooks and programs (software);
- suitable training courses;
- system compatibility and expansion capability.

The technical facilities of the system which will be governed to some degree by the system supplier, are:

- input/output capacity;
- type of input/output, i.e. analog, digital, PID, etc.;
- memory size (storage capacity);
- scan rate (speed of the system);
- type of instructions available in the instruction set;
- service backup.

The input/output requirements must be specified in terms of the number of digital input points and digital output points required, the voltage levels, AC or DC, any special input/output assemblies such as servo motor controllers, analog, etc., and user power supply requirements. If a modular system is being used, sufficient rack space must be available to facilitate additional modules for future expansion.

The memory storage capacity requirements will vary depending on the complexity of the user program and the number of the I/O connected to the system. The memory space that complex functions take up, such as block storage, as well as future additions or alterations to the system, must be also considered and be provided for. A study of the manufacturer's memory allocation chart will assist in estimating the memory storage capacity and its suitability for the user's requirements. The memory of most programmable controllers can be expanded by including additional memory modules or adding additional memory chips.

The program scan rate is an important consideration where control of high speed functions is necessary. The scan rate is directly related to the speed at which the program functions are executed. The time taken to scan 1 k of memory can range from

0.9 microseconds to 60 milliseconds. Immediate, or interrupt instructions, or high speed program execution areas in the program, are available in programmable controllers to obtain the necessary high speed execution of program functions.

The instruction set of the programmable controller must have all the instructions available that are necessary to perform the control task. If the control functions have been clearly specified, the range of instructions required for the program will be apparent. Special functions, such as complex arithmetic, PID control, and data communication facilities, should also be considered. Some consideration must be given to the availability of PLC manufacturer software which will allow a personal computer to be used as PLC programmer. This would either eliminate the need for a costly, dedicated PLC programmer or terminal, or reduce the dedicated programmer need to a simple hand-held terminal. Most programmable controller manufacturers have software available to allow PLC programs to be designed on a personal computer, but a cost analysis must be done of the software required to perform the programming function and this must be compared to the cost of a dedicated program terminal in order to determine the best option for the user.

13.3 Field wiring and terminations

Warning: All wiring connected to a programmable controller and associated machine or process must be carried out in compliance with the standards set by the wiring regulatory authorities.

Mounting the programmable controller

Consideration should be given to the location in which the programmable controller is to be mounted. Locations which may suffer vibration or mechanical shock can lead to the printed circuit board becoming dislodged or component leads breaking away from the printed circuit board. High temperature should also be avoided.

Although the programmable controller has been designed to operate in an industrial environment, the device must be located away from:

- excessive temperature;
- excessive humidity;
- corrosive atmospheres;
- metal laden oils;
- conductive dust.

Exposure to any of these harsh industrial environments could result in the failure of solid state control devices. If the programmable controller is to be located in a harsh environment it must be in a contaminant-proof enclosure.

Electrical noise

The operation of solid state control equipment can be seriously affected by electrical noise being induced into control circuit wiring. Electrical noise is defined as any unwanted electrical signal produced by a magnet or electrostatic field that is induced into the control circuit by various means. Induced noise often produces intermittent equipment malfunctions because it is random. Programmable controller input circuitry will respond to the short-duration induced noise because the solid state input components have a fast response time. Input noise filters and contact debounce circuits will reduce the effect of the noise, but it is preferable to provide effective shielding of the field cable to reduce the amplitude of the induced noise. Figure 13.1 shows a twisted pair field cable which is shielded. It provides induced magnetic field immunity due to the 180° phase reversals produced by the twisting of the pair. Electrostatic shielding is also provided by the shield around the cable which is connected to earth at one point only. Grounding one end of the shield will avoid ground loops where grounds of different potential cause current to flow through the shield. Ground loops can induce 'hum' into the input circuits.

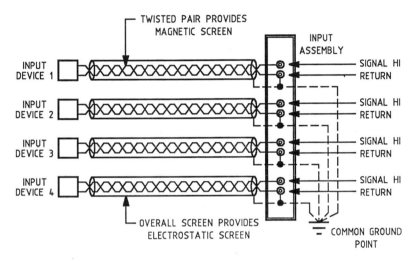

Figure 13.1   Field cable connections

Grounding programmable controllers

The correct grounding or earthing procedure for a programmable controller will improve noise immunity, is an important safety requirement, and must comply with electrical codes. A guideline for grounding requirements in the USA, titled the National Electrical Code, can be obtained from the National Fire Protection Association of Boston, Massachusetts. Although national codes are available, it is important that local and state electrical codes be consulted as electrical codes may vary between locations. The grounding path must always be:

- permanent;
- continuous;
- of sufficient size to be able to conduct fault current;
- of minimum resistance.

A ground lug will be provided on the programmable controller rack or frame which must be connected to a correctly grounded, ground bus connection, as shown in Figure 13.2.

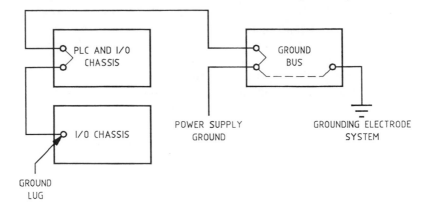

Figure 13.2   Typical grounding connection

13.4 User power supplies

The power supply connections, both AC and DC, to the programmable controller and controlled process or machine are an important consideration, as the safe operation of the system with respect to personnel and equipment, as well as the system reliability, can be improved by a well-designed power distribution system. Figure 13.3 shows a power distribution configuration suitable for a programmable controller installation.

The constant voltage isolation transformer provides:

- physical isolation from the main power distribution;
- constant secondary voltage to reduce the effect of main power distribution fluctuations;
- voltage transformation, if required, to provide 110 or 240 volts to the AC distribution system.

The master control relay (MCR) network is a standard start/stop configuration. Hardwired emergency stop switches are used to stop the PLC system by releasing the master control relay. The switches would be located at appropriate positions throughout the PLC and controlled equipment installation. As the emergency stop switches provide a safer working environment for personnel and controlled equipment, any number of them can be wired in series. The emergency stop switches are totally independent of the rest of the controlled system. The contacts of the master control relay disconnect power to the AC and DC input/output assemblies.

The sizing of the equipment used in the PLC power distribution system must be considered. Sizing of equipment relates to the current, voltage, and power ratings of the controlled devices, and therefore the power supplies. For example, the DC power

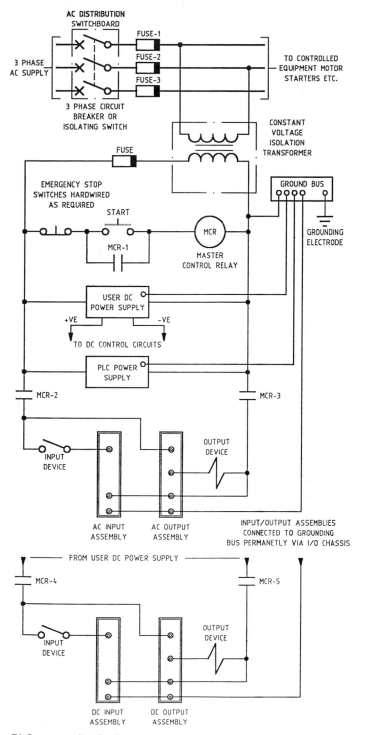

Figure 13.3   PLC power distribution system

supply must be able to provide the current required to operate all outputs simultaneously without a significant drop in supply voltage, even though it is unlikely to have all outputs on together. Sizing the DC power supply using maximum output current will stop the rating of the power supply from being exceeded, and would allow for some future system expansion without changing the power supply. It is important to note that the power supply current rating is directly related to the number of DC input and output devices being supplied or driven, and that the current requirements of 24 volt DC control circuitry is often high. The power requirements for each device must be known before accurate power supply sizing can be achieved.

13.5 System documentation

The operation of a programmable controlled system is dependent on the clarity and ease of use of the system documentation. The system must be accurately documented from conception to completion, and, although documentation costs may appear to be high, down time will be greatly reduced when troubleshooting, servicing or commissioning, if the system is documented clearly and accurately. In addition, future system modifications can be carried out more easily if the original documentation is continually updated and contains all subsequent modifications. Incorporating the system documentation into training programs is also an advantage when introducing new personnel to the system, or reviewing the system with existing maintenance personnel. Well laid out documented system information, used on a regular basis for maintenance, becomes a major servicing and troubleshooting tool. An accurately documented system should contain the following information:

- a written description of the controlled machine or process;
- a block diagram of the complete system;
- a list of the objectives being met by the controlled system;
- an explanation of the role of the programmable controller in the controlled process, including interaction with other equipment or processes;
- the manufacturer's programmable controller handbooks;
- flow charts of the user program;
- a program listing, cross referenced between rungs and including an operational description;
- software listings including programs used in EPROMS;
- wiring schedules, including cable numbers and wire ferrules;
- a description of the connected peripherals;
- the operator's manual, including start up and shutdown procedures;
- an alarm checklist, with details of what to do when an alarm occurs, and possible causes of alarms;
- memory configuration charts (memory maps).

The use of personal computers and programmable controller manufacturer software allows quality documentation to be produced as the ladder diagrams and sequential functions are designed. Documentation produced by using a personal computer to design the ladder logic will usually contain rung cross references where repeatedly used addresses are cross referenced to rung numbers, availability of unused internal and external storage points, and memory configuration charts. Figure 13.4 is a sample of a PLC ladder diagram documentation available from a personal computer. Using the personal computer to design the program also allows the program to be designed away from the PLC, stored on a mass storage system such as a floppy disk, and loaded into the programmable controller via a data communications network, or directly into the programmable controller memory as required.

13.6 Installing and commissioning PLC systems

The installation of a PLC system commences with the layout of cable routing to the chosen location of a suitable cubicle to house the programmable controller. If the machine or process equipment is already located, cable routing options are somewhat limited, but it is still possible to design the cable route to obtain the shortest possible connection distance between the controlled device and the programmable controller. Short cable routes reduce cost and are less likely to be susceptible to induced noise. The selection of cable is also important, and, where possible, the field cable should be shielded, twisted pairs. If necessary, armored cable can be a long-term cost effective option; it is essential in potentially dangerous environments such as those where processes involving explosive or flammable materials are used. All field cabling should enter and exit the PLC cubicle via a dust-proof and moisture-proof cable gland. The area within the PLC cubicle is a clean area where unshielded interconnection cables are likely to be found, such as ribbon cables. If screw terminals are provided for I/O connection of field cables, the correct size crimp-on lug should be used to terminate the cable otherwise loose or intermittent cable termination may result.

Each wire in the cubicle should have an identifying ferrule. The cables should be grouped, labeled, and separated according to type, i.e. AC signal, DC signal, and power cabling should be kept separate and clearly identified. Grouping of cables will assist in reducing system noise. The cubicle housing the PLC will vary depending on the environment in which the PLC is to be located. It is a minimum requirement to house the PLC in a dust-proof and moisture-proof cubicle, however, if necessary, consideration should be given to corrosion and fume proofing. The cubicle should be lockable with a viewing area to allow operators to view the PLC status indicators without opening the cubicle. The installation of door alarms and safety interlocks for the cubicle doors is also an option for consideration, as is the choice of PLC control switches and keys which would be mounted into the cubicle door to allow for operator access. The PLC cubicle must also be grounded according to the required standards.

Commissioning PLC systems

Commissioning a PLC controlled system is an important procedure as incorrect installation or a program fault can injure personnel or cause damage to the controlled equipment. The following steps are taken when commissioning the system while the PLC is in the test (non operating) mode and the controlled plant is decommissioned:

1 Check each cable connection and the continuity between the PLC I/O terminals and the field devices.

```
01-01-1980                                          ANNUNCIATOR ASSIGNMENT
00:03:45                                            Ladder Diagram Listing
-----------------------------------------------------BROADMEADOWS COLLEGE OF TAFE----------------------
    ¦Tmr2(dly)                                              Timer 2¦
    ¦ 031/15                                                   031 ¦ [15/] 1.
  1 +---]/[------------------------------------------------------(Ton)-+ [17/]
    ¦(1)   D                                                    0.1 ¦ [Word   031] 9.
    ¦                                                        Pre 040 ¦
    ¦                                                        Acc 029 ¦
    ¦ Timer 1                                                Tmr 1(2s)
    ¦ 030/15                                                    030 ¦ [15/] 2.
  2 +---]/[------------------------------------------------------(Ton)-+ [17/]
    ¦(2)   D                                                    0.1 ¦ [Word   030] 3.
    ¦                                                        Pre 020 ¦
    ¦                                                        Acc 009 ¦
    ¦    030       042      060/02                          020/00¦ -] [- 5.
  3 +--[Get]--+---[<]---+---]/[-----------------------------------( )--+
    ¦  009(2)¦   010  ¦(8)   S                                     S¦
    ¦          ¦  060/01 ¦                                          ¦
    ¦          +---] [---+                                          ¦
    ¦             (7)   S                                           ¦
    ¦ alarm 1                                                      ¦
    ¦ 110/00                                                060/00¦ -(U)- 47.
  4 +---] [------------------------------------------------------(L)--+ -] [- 5. 8. 46.
    ¦    S                                                          S¦
    ¦ 020/00   060/00                                        020/02¦ -] [- 6.
  5 +---] [---+---] [-----------------------------------------( )--+
    ¦(3)   S ¦(4)   S                                            S¦
    ¦ 020/01 ¦                                                     ¦
    +---] [---+                                                    ¦
    ¦(9)   S                                                       ¦
    ¦ 020/02                                                011/00¦
```

Figure 13.4   PLC sample documentation

```
6 +---] [---+--------------------------------------------------------( )--+
  |(5)   S |                                                          S|
  | 110/16 |                                                           |
  +---] [---+                                                          |
  |      S                                                             |
  |TEST P/B                                                            |
  | 110/10                                                 060/01| -]/[- 46.
7 +---] [------------------------------------------------------------(L)--+ -(U)- 48.
  |      S                                                   S| -] [- 3.
  | alarm 1                                                    |
  | 110/00    060/00                                       060/02| -]/[- 3.
8 +---]/[-------] [----------------------------------------------------(L)--+ -(U)- 49.
  |     S  (4)   S                                           S| -] [- 9.
  | 060/02     031      043                                020/01| -] [- 5.
9 +---] [------[Get]------[<]----------------------------------------( )--+
  |(8)   S    029(1)   020                                   S|
  | 033/15                                                  033 | [15/]  10.
10 +---]/[------------------------------------------------------------(Ton)-+ [Word   033] 18.
  |(10)   D                                                 0.1 |
  |                                                       Pre 040 |
  |                                                       Acc 029 |
  | 032/15                                                  032 | [15/]  11.
11 +---]/[------------------------------------------------------------(Ton)-+ [Word   032] 12.
  |(11)   D                                                 0.1 |
  |                                                       Pre 020 |
  |                                                       Acc 009 |
  |   032      044      060/05                            020/03| -] [- 14.
12 +--[Get]--+---[<]---+---]/[----------------------------------------( )--+
  | 009(11|   010  |(17)  S                                 S|
  |        | 060/04 |                                         |
  |        +---] [---+                                         |
  |         (16)  S                                           |
```

Figure 13.4   PLC sample documentation (cont)

2 Test and verify the operation of the emergency stop facility.

3 Manually operate all field devices and check that they are connected to the correct I/O, and that they are in the correct state, i.e. normally open, normally closed, etc.

4 Simulate the signals for each analog loop.

5 Force each output 'on' or 'off' in turn to ensure the correct operation of the field devices. The consequences of each force operation must be considered prior to forcing each output.

6 When the field devices and wiring have been thoroughly checked against the documentation, software and control simulation can commence.

The results of each test should be documented.

As most programmable controllers have software checking facilities, the program, once installed, can be checked for syntax (programming language) errors, multiple labels, incorrect jumps, etc. Simulation of the controlled process or machine under the control of the PLC program can in some instances be carried out. In large or complex control applications, simulations may be difficult, but it is worthwhile verifying the program; full simulation using switches, lamps, and signal generators may be well worth the effort. The use of the program monitor facility to indicate the status of program rungs or individual inputs or outputs is a useful method of determining if the actual and expected operation are the same.

The final step in the commissioning procedure is to install and run the user program. A well-designed program which allows sections of the program to be run and tested independently of other program sections will assist in commissioning troubleshooting. If the program has not been designed in sections the insertion of program jumps and temporary end-of-program commands can break the program into sections.

A programmable controller which allows single stepping of the program, in conjunction with the monitor facility, will indicate whether the operational sequence of the program is correct.

The final test is to actually run the system. The system should be observed, and, if unexpected operations occur, the emergency stop should be used to stop the process or machine and the commissioning procedure should be repeated.

13.7 Test questions

Multiple choice

1 When carrying out a programmable controller needs analysis which of the following would not be a consideration?

(a) Technical requirements.

(b) Economic impact.

(c) Integration into the workplace.

(d) PLC instruction set.

2 Which of the following technical facilities must be considered when purchasing a programmable controller?

(a) I/O capacity.

(b) Grounding requirements.

(c) Scan rate.

(d) Sequence control.

3 An installed and operational programmable controller intermittently fails due to printed circuit board short circuits between tracks. Which of the following is the most likely cause?

(a) Excessive temperature.

(b) Excessive humidity.

(c) Corrosive atmosphere.

(d) Conductive dust.

4 Which of the following are requirements of a ground or earth system for a programmable controller?

(a) Minimum resistance to ground.

(b) Continuous connection.

(c) Aluminum ground wire.

(d) Permanent connection.

5 Emergency stop switches must always be:

(a) hardwired.

(b) in momentary operation (non latching).

(c) green.

(d) in a break glass enclosure.

True or false

6 Personnel training is an unnecessary requirement when using a programmable controller for an industrial application.

7 The scan rate of a programmable controller will vary from 0.9 microseconds up to 60 milliseconds per 1 k of memory.

8 Personal computers, with the appropriate software, can be used as PLC programmers.

9 The shield of PLC field wiring must be earthed at both ends to provide earth loops.

10 The cubicle in which the PLC is mounted must always be grounded to enhance safety and reduce noise.

11 One emergency stop switch only is to be wired into a programmable controller system.

Written response

12 What effect does twisting the field wiring into pairs and shielding the pairs have on electrical noise reduction?

13 Briefly describe the functions of the isolation transformer used in the AC distribution system for a programmable controller.

14 Describe a logical commissioning procedure for a PLC system and briefly explain the selected order of commissioning procedure.

15 What consideration must be given to sizing of a PLC DC user power supply?

16 List five documentation requirements that should be supplied during PLC system design.

Troubleshooting in PLC systems

14.1 PLC status indicators and alarms

The status indicators provided with programmable controllers fall into one of three categories:

- operation and program flow indicators;
- operator response indicators;
- technical response indicators.

The operation and program flow indicators, such as the run, DC and AC power on indicators, require no action and indicate that the programmable controller is functioning correctly. When these indicators extinguish, some action is usually required by the operator. Program flow indicators can be provided to the operator in the form of a mimic panel, mimic diagram on a computer, or graphically. As the operator is on hand it is usual for that person to have overriding control and responsibility for the controlled process or machine, and therefore a good knowledge of the programmable controller and a detailed knowledge of the controlled process or machine. The type of input that the operator requires to have into the system includes:

- alarm acknowledgement;
- system shutdown, halt, or pause;
- preset, or change of system parameters;
- inhibit sections of the system;
- change system programs.

Some indicators require operator response to gain technical assistance, but the indicators that are not urgent, termed maintenance indicators, define the status of the system. One such maintenance indicator is standby battery low for memory back-up, which, when on, would prompt the operator to advise the maintenance department to schedule the required service.

The indicators that require technical response are those which cause disruption to the controlled machine or process, such as power supply fail indication. These indicators require immediate attention. When mimic or advisor displays are used to assist with alarm annunciation, the operator input is via a standard computer keyboard which can be located some distance from the controlled process. Data communication techniques are used to gather the required information. Many programmable controller manufacturers provide the software necessary to allow mimic display generation.

The status indicators located on the programmable controller are an indication that problems have occurred within the programmable controller. The status indicators will vary between PLC manufacturers, but most have the following:

- operational mode;
- processor fault;
- memory fault;
- I/O fault;
- I/O on/off;
- DC/AC power on;
- standby battery low;
- communications failure.

The selected mode of operation of the PLC processor, i.e. run, test, program, halt, etc. is indicated by the select switch position and associated indicator. The mode indicator provides information to the operator to assist when program changes and their housekeeping functions are being undertaken during the normal operation of the system.

The processor fault indicator, when 'on', indicates that a major fault has occurred in the processor module. Cycling the power (turning power off, then back on) may clear this fault, but continuing failures will require processor module replacement.

The memory fault indicator, when 'on', indicates that there is reasonable doubt that the contents of the memory are correct. The memory fault indicator is usually generated by parity checks. Reloading the memory will often clear this problem, but the reason for the memory data being corrupt should be investigated. Continuing problems of corrupt memory information may require replacement of the memory modules.

The I/O fault indicator indicates a communication error between the processor and I/O rack. All interconnection communication cables should be checked. Some I/O fault indicators show when fuses in the I/O have blown due to field faults; they serve to assist when troubleshooting faulty I/O. The I/O on/off status LEDs indicate when a field input has closed and the input is in an active state. The output indicator is 'on' when an output signal is available at the PLC output terminals to operate an output field device.

Power supply AC on indicator, when 'on', indicates that AC power is connected to the power supply. The DC power LED when 'on' indicates that DC power is available for the processor and I/O assemblies. If the AC and DC power indicators are not lit, the AC supply and associated fuses should be checked.

The standby battery low indicator, when 'on', indicates that the battery backup for the volatile memory during mains failures should be replaced. The standby battery low LED does not indicate a PLC failure but indicates that battery maintenance is required.

Error messages

Programmable controllers are often designed to display error codes or error messages when a fault causes the PLC to fail. Error code summary sheets are provided with the manufacturer's instruction manuals and should be consulted when error codes occur.

Testing input/output assemblies

The input assembly LED associated with each field input connection is an excellent indication that the field device is active or inactive, although in some instances it may

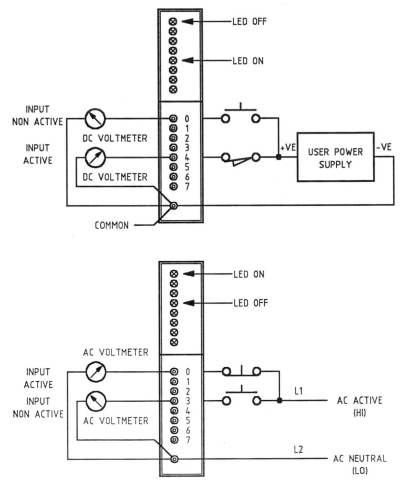

Figure 14.1(a & b)   Testing input assemblies

be necessary to verify the operation of an input using a voltmeter. The DC inputs shown in Figure 14.1(a) can be checked by placing the voltmeter on the appropriate DC range,and connecting the negative lead of the meter to the common and the positive lead of the meter to the input under test. If the meter reads a value equal to the user power supply voltage, the field device is providing a closed circuit; no reading on the meter indicates the field device is open. The AC input assembly shown in Figure 14.1(b), is tested in a similar manner using an AC voltmeter.

Testing of output assemblies can be more complex, particularly if the load is disconnected or field wiring open circuit. To check the operation of the output assembly it is advisable to place a load resistor across the output during the test. The load resistor is required as the high impedance of the multimeter may not cause the assembly output solid state device to conduct, and an incorrect voltmeter reading may result. Figure 14.2 shows the test connection of DC and AC output assemblies respectively, using a voltmeter.

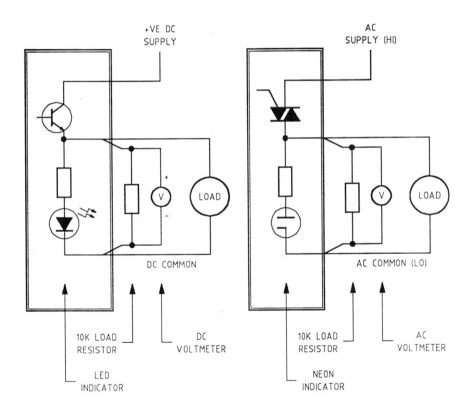

Figure 14.2   Testing output assemblies

14.2 Using the force instruction

The 'force on' and 'force off' instructions may be used to assist in system trouble-shooting or commissioning. These instructions can be used to operate or release an output, therefore checking the complete electrical circuit from the output assembly. When a force instruction is used on an output, the output will assume the forced condition regardless of the condition logic in the rung, and will remain in the forced state until the force instruction is removed.

Warning: Using the force instruction without a complete understanding of the consequences of forcing a data table bit on or off could cause injury to personnel and/or hazardous machine operation.

14.3 Troubleshooting flow charts

The use of troubleshooting flowcharts makes fault location in a PLC controlled system considerably easier and less time consuming. Not all PLC manufacturers supply troubleshooting flowcharts, but one specific to the installed system could be designed as part of the overall system documentation. Troubleshooting flowcharts also assist maintenance personnel in identifying areas of responsibility. A simple troubleshooting flowchart which could be used to assist an operator to carry out a preliminary check of the PLC prior to calling in the maintenance technician is shown in Figure 14.3. The chart could be expanded to cover a range of possible faults, and would be of use to the maintenance technician.

14.4 Identifying PLC system faults

As the systems in which programmable controllers are the controlling device become more complex, it becomes increasingly difficult to identify programmable controller system faults. The programmable controlled system can be divided into eight broad areas in which faults can be classified. These areas are shown in Table 14.1 with an approximate percentage of the total faults associated with each area.

Table 14.1   Fault distribution in a PLC system

Fault catagories	% of total faults
1 Field devices digital	33
2 Field devices analog	20
3 Communications devices	5
4 Field wiring	10
5 Induced fault conditions	5
6 PLC processor unit	5
7 PLC I/O units	14
8 PLC power supply	6
8 Other	2

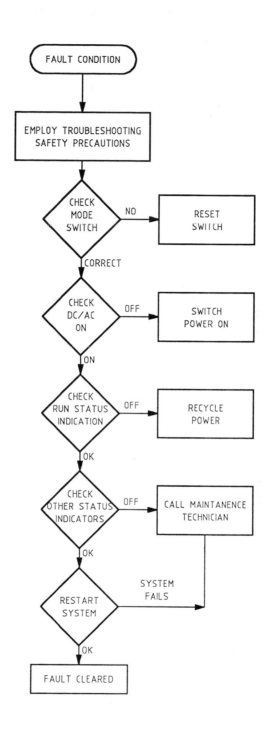

Figure 14.3 Troubleshooting flowchart

Categories 1 to 5 in Table 14.1, which comprise 73 per cent of the total faults in the system, are faults which are a result of components external to the programmable controller. Of the remaining 27 per cent of the total system faults, 14 per cent are associated with I/O units, of which blown fuses contribute the major portion, and are likely to be as a result of work being carried out by maintenance or installation personnel associated with field wiring. The percentage of faults in Table 14.1 has been derived from fault occurrence information from a number of systems. The fault totals include post system commissioning fault occurrence, which would normally have higher fault numbers than long-term post commissioning fault occurrence. It was also noted when gathering fault information that many system users did not keep accurate fault occurrence records.

The fault occurrence in PLC systems follows, in general, the normal distribution curve with respect to time as shown in Figure 14.4. The post commissioning fault occurrence is high, but quickly falls to a constant rate of failure which tends to increase toward the end of the considered useful life of the PLC system. The fault occurrence in the PLC does not increase significantly with time, but experience has shown that field devices tend to be the main source of overall system failures toward the end of the useful life of the system. Programmable controllers are not usually replaced during the useful life of the machine or process and are decommissioned only when the total system (process or machine) becomes obsolete.

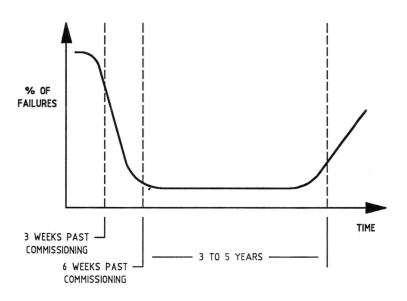

Figure 14.4   Fault occurrence in PLC systems

14.5 System troubleshooting techniques

A systematic and logical approach must be used to troubleshoot a PLC controlled system. In the event of a system failure, the steps to be taken to correct the problem are:

1 Consider all symptoms.
2 Make the system safe for troubleshooting.
3 Localize the problem.
4 Isolate the fault.
5 Carry out corrective action.
6 Reinstate or recommission the system.

When the system fails, consideration of the system as a complete unit is essential, i.e. has the system completely ceased to operate, or is the system part operating? Observe diagnostic indicators and alarms. Make the system safe to continue to troubleshoot by tagging with safety tags where appropriate. Advise operators, erect warning barriers, or take any other steps necessary to ensure plant personnel are not injured or machinery is not damaged during troubleshooting. System safety is of prime consideration when working with automatic machines or processes.

Localizing the problem into an area may be as simple as determining if the problem is in the field or in the programmable controller in a basic system, or determining which rack of equipment is not functioning correctly in a complex system.

The isolation of the fault to a single field unit or one module in the system then requires the replacement of the faulty unit. This must be carried out while the system has the supply power switched off. When the fault has been cleared, the system must be reinstated or recommissioned using the same safety precautions used when the system was commissioned. The faulty module or unit of equipment must then be repaired or replaced, so that spare part stocks are maintained.

Use of the program terminal to monitor the operation of rungs will also greatly assist in determining correct operation of the user program, and locating field devices that are operating out of sequence or field devices which may have failed.

14.6 PLC maintenance techniques

The programmable controller system should be included in and be part of a preventative maintenance program which will decrease the down time of the system due to breakdown. The following checks, approximately once a month, should be performed regularly:

1 Visually check all field input and output devices for damage, wear, and physical location.
2 Check the temperature at the PLC, and inside the PLC cubicle, as high temperature may cause the PLC to fail.
3 Avoid dust and other forms of environment contamination by cleaning the PLC enclosure.

4 Check the mechanical rigidity of all terminal strips and connectors.

5 Consider the overall system for sources of electrical noise, including system grounding.

6 Check backup software for validity and obsoleteness.

7 Check stocks of fuses and other expendable items. If the system is a modular type, a full set of spare modules should be retained. A small PLC system may have a spare programmable controller in stock as a unit replacement.

Warning: Damage to the system or haphazard operation of the machine or process may occur if fuses or modules are replaced while power is applied to the system.

8 Log dates, times, and symptoms of any faults that occur in the system for future reference and maintenance records.

9 Training of operating and maintenance personnel should be carried out on the system. The high reliability of programmable controllers causes problems for maintenance personnel because the long duration between faults means refresher training on the system is necessary.

10 Make sure system documentation is updated and available.

14.7 Test questions

Multiple choice

1 Testing between the input terminal and common of a DC input module when the field wiring is short circuited at the field device terminals will cause the DC voltmeter to read:
   (a) the user power supply voltage.
   (b) half user power supply voltage.
   (c) zero.
   (d) 24 volts DC.

2 The force instructions:
   (a) will force data table bits 'on' only.
   (b) will force data table bits 'off' only.
   (c) if used indiscriminately, could cause haphazard machine operation.
   (d) can only be used to force inputs.

3 Most PLC system faults occur in:
   (a) field devices.
   (b) processor units.
   (c) input/output units.
   (d) field wiring.

4 The prime consideration when troubleshooting PLC systems is:
   (a) to minimize down time.
   (b) system safety.

(c) fault isolation.

(d) observing the symptoms.

5 To test an AC output assembly, the AC voltmeter must be placed between the:

(a) AC common or neutral and the output terminal with a 10 k load resistor in parallel.

(b) AC active or the output terminal with a 10 k load resistor in parallel.

(c) AC common or neutral and the active or Hi with a 10 k load resistor.

(d) the AC common terminal or neutral and in parallel with the field load.

True or false

6 Troubleshooting flow charts make fault location easier, but increase fault location time due to the complexity of the charts.

7 PLC modules should be inserted or removed while power is connected to the system.

8 A systematic and logical approach should be used when PLC system troubleshooting.

9 The useful life expectancy of a PLC controlled system is three to five years or the useful life expectancy of the controlled machine or process.

10 Preventative maintenance increases down time.

Written response

11 Briefly describe the function of error messages.

12 List the status indicators commonly found on programmable controllers and briefly describe the function of each.

13 List the logical steps to be taken when troubleshooting a programmable controller.

14 List the requirements to establish a preventative maintenance procedure for a PLC system.

15 Describe how a PLC program terminal can be used to assist in PLC system troubleshooting.

Appendix 1

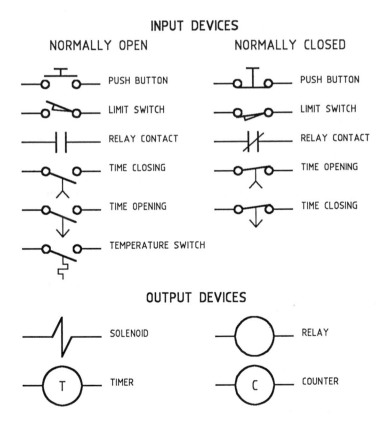

INPUT DEVICES

NORMALLY OPEN   NORMALLY CLOSED

PUSH BUTTON   PUSH BUTTON

LIMIT SWITCH   LIMIT SWITCH

RELAY CONTACT   RELAY CONTACT

TIME CLOSING   TIME OPENING

TIME OPENING   TIME CLOSING

TEMPERATURE SWITCH

OUTPUT DEVICES

SOLENOID   RELAY

TIMER   COUNTER

Appendix 2

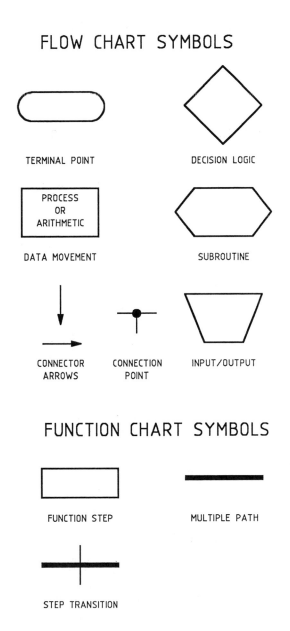

FLOW CHART SYMBOLS

TERMINAL POINT

DECISION LOGIC

PROCESS
OR
ARITHMETIC

DATA MOVEMENT

SUBROUTINE

CONNECTOR
ARROWS

CONNECTION
POINT

INPUT/OUTPUT

FUNCTION CHART SYMBOLS

FUNCTION STEP

MULTIPLE PATH

STEP TRANSITION

Appendix 3

This section provides answers to all odd-numbered questions. Note that multiple choice questions may have more than one correct answer.

Chapter 1

Multiple choice

1. (b) hardwired electronic control
3. (b) jump
5. (d) 252
7. (d) 883
9. (d) 362
11. (d) last in first out (LIFO)

Written response

13. The advantages of a programmable controlled system as compared to a relay system are:
- System function can be modified or changed by reprogramming without wiring alterations.
- The PLC is smaller in size and therefore take's up less rack space.
- The programmable controller can produce its own documentation.
- Counters, timers, sequences and many other functions are available.
- No special PLC programming knowledge is required.

15. Flags are used in the microprocessor to provide signals when decisions within the system are to be made.

Chapter 2

Multiple choice

1. (b) The time it takes for the contact to operate
   (c) The electrical interference produced
   (d) The amount of current that can be passed through the contact
3. (b) random access memory (RAM)
5. (a) input/output assemblies
   (b) programming unit
   (e) processor
7. (c) convert the input AC to DC
9. (b) the output is active

True or false

11. False
13. False
15. False

Written response

17. Bits are 1s and 0s which represent the program in the memory of a programmable controller. The 1s and 0s can be in the form of voltage levels, magnetic fields or magnetic field polarity. The word 'bit' is derived from two words binary and digit.

19. The two power supplies found in PLC systems are the :

- internal or local power supply
- user or remote power supply

The internal power supply provides DC power to operate the processor, input/output modules, and any other equipment in the PLC rack.

The user power supply may be AC, DC or both AC and DC. It supplies power for the input and output load devices in the system. In some PLCs the user power supply is separate from the rest of the PLC system.

Chapter 3

Written response

1.

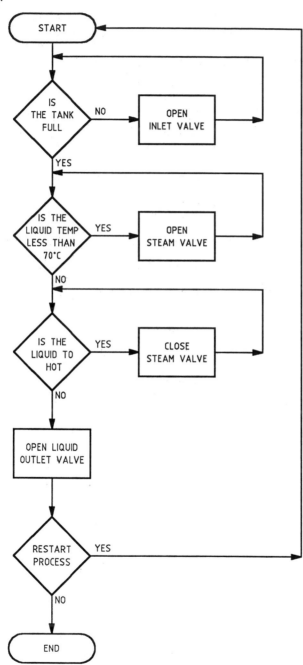

3. Ladder diagram for Figure 3.28

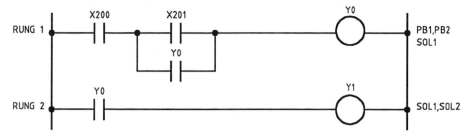

KEY STROKES HITACHI P250-E

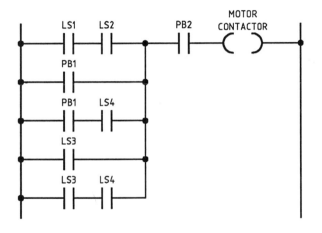

5. Ladder diagram for Figure 3.30

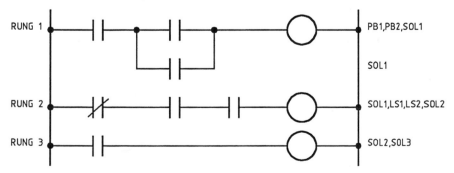

7. Decision chart for Figure 3.32 Part 1

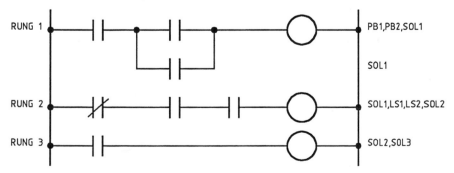

269

7. Decision chart for Figure 3.32 Part 2

OUTPUT DEVICE OPERATION	OPERATION NUMBER	INPUT/OUTPUT CONDITION							RUNG NUMBER
		PUSHBUTTON 1	PUSHBUTTON 2	SOLENOID 1	LIMIT SWITCH 1	LIMIT SWITCH 2	SOLENOID 2	SOLENOID 3	
SOLENOID 1	1	Ō	O						1
	2	Ō		O					
SOLENOID 2	3			Ō	O	O	O		2
SOLENOID 3	4						O	O	3

Chapter 4

Multiple choice

1. (a) program
   (c) write
   (e) edit.
3. (a) analog input.
   (b) digital input
5. (d) user program input and output scan time.

True or false

7. False
9. True

Written response

11 (a)   AND function ladder diagram

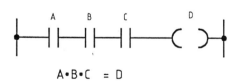

$$A \cdot B \cdot C = D$$

A	B	C	D
0	0	0	0
0	0	1	0
0	1	0	0
0	1	1	0
1	0	0	0
1	0	1	0
1	1	0	0
1	1	1	1

AND LADDER DIAGRAM   TRUTH TABLE

11 (b)   OR function ladder diagram

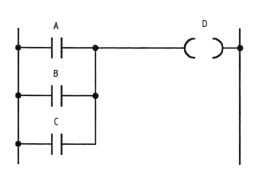

$$A+B+C = D$$

A	B	C	D
0	0	0	0
0	0	1	1
0	1	0	1
0	1	1	1
1	0	0	1
1	0	1	1
1	1	0	1
1	1	1	1

OR LADDER DIAGRAM   TRUTH TABLE

13.
1. Jumping counters and timers will stop them from being incremented. Programming the counter and timer outside the jumped area can overcome this problem.
2. Jumping to no accessible data table locations may cause the program scan to lose its location.
3. Jumping backward or to an undefined label may result in an unpredictable scan sequence.
4. Jumping to a destination within a master or zone control area could result in unpredictable machine or process operation.

15. Ladder diagram

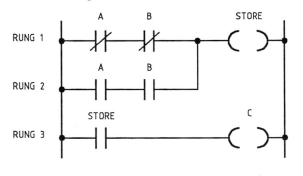

A	B	STORE	C
0	0	1	0
0	1	0	1
1	0	0	1
1	1	1	0

LADDER DIAGRAM   TRUTH TABLE

271

Chapter 5

Multiple choice

1. (c) to time a period between maximum and minimum
3. (c) remain the same
5. (d) the time taken to scan input, outputs and user program

True or false

7. False
9. True

Written response

11. The counter is reset each time the acculated is equal to the preset valve 04015 or the counter underflows or overflows 04014. (Program is for the Allen Bradley PLC 2).

Ladder diagram for an up/down counter

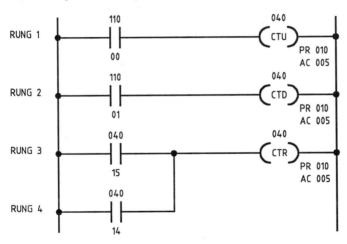

13. Timing diagram

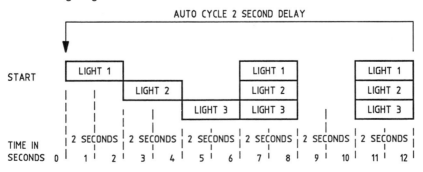

In this application use is made of the accumulated BCD bits which are accessible in an Allen-Bradley PLC 2.

Ladder diagram

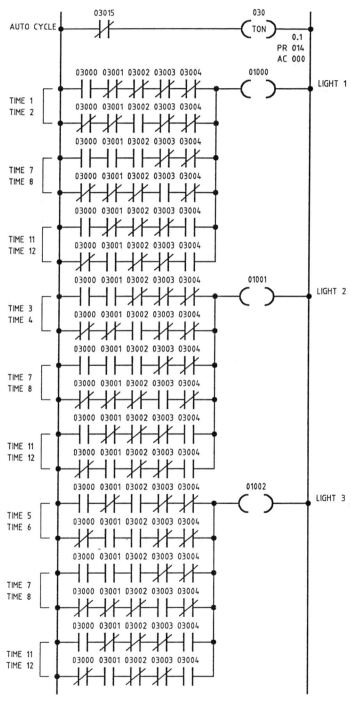

15. After 5000 one-second pulses output Y0 will be latched.

Ladder diagram

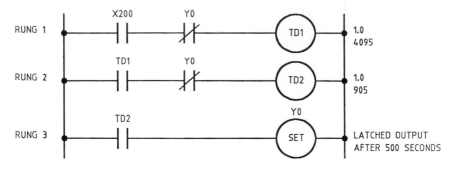

Chapter 6

Multiple choice

1. (b) 0011 0100 0011 0000
3. (d) carry out complex arithmetic
5. (a) load counter and timers with accumulated values
   (c) compare functions
   (e) arithmetic functions

True or false

7. False
9. True

Written response

11. The fixed value is being compared against word 020 variable value in word 110.

Ladder diagram

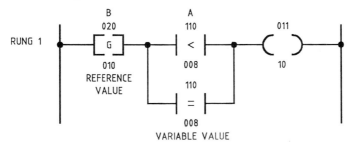

13.

Ladder diagram

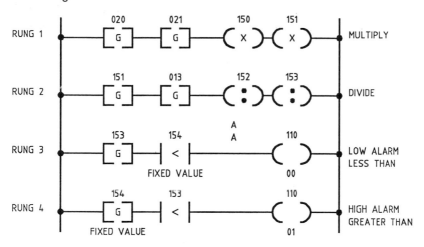

15.

Ladder diagram

Block diagram

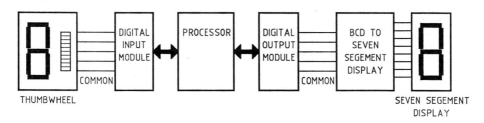

Chapter 7

Multiple choice

1. (d) shaft position
3. (b) the smallest detectable input change that the instrument is able to detect
5. (a) open and closed loop

True or false

7. True
9. True

Written response

11. Batch process has set, identifiable, discrete loads of raw material which are sequentially introduced into the process. At the conclusion of the batch process the material may be further processed or stored. Applications of batch processes are the manufacture of glue, paint, etc. where chemicals are being mixed.

Continuous process has raw materials continuously fed into the system; finished product is the output. The raw materials may not be able to be individually tracked through the system. The plastic industry blowmoulding process is an example of a continuous process.

13. A current analog signal is less susceptible to electromagnetic or electrostatic interference in the form of noise.

The voltage drop over the conductors of an analog voltage signal may introduce an unwanted error.

The 4 mA signal being equal to 0 or the lowest analog value can be used to detect a fault condition when the circuit current falls below 4 mA. Zero volts in a 0 – 5 volts signal is not a fault condition. If the circuit was open zero volts would result.

15. Block diagram data acquisition system

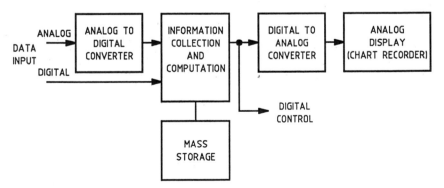

1. Analog-to-digital converter accepts analogs as an input and converts the analog to its digital equivalent.
2. Information collection and computation accepts digital inputs and analog converted inputs and carries out the required computation on the input data under the control of the user program. The processed data is fed out as digital signals.

3. Mass storage is memory where the user program computer results, data tables, etc. are stored.

4. Digital-to-analog converter converts the digital outputs to their analog equivalent.

5. Analog display, such as a chart recorder, displays the analog signal.

Chapter 8

Multiple choice

1. (b) program mode
3. (d) register
5. (d) stack

True or false

7. True
9. True – with appropriate software

Written response

11. The consequences of on-line programming functions must be considered before they are carried out. On-line changes may cause machinery or processes to operate haphazardly and possibly endanger personnel or damage machinery.

13. A file-to-file move is where a group of data table words set up as a file containing data and defined start and end has the bit pattern of the file transferred to a new data table address location in the data organization table.

Chapter 9

Multiple choice

1. (b) words

3. (d) sequencers

5. (b) true

True or false

7. False
9. False

Written response

11. The function of the mask in the sequencer is to allow various bits, in the step of the sequence being scanned, to be turned 'ON' or 'OFF' depending on the bit configuration of the mask. This is achieved by allowing outputs to occur if the mask bit is set to a '1' and the word bit being scanned is '1'. If the scanned bit is '1' and the mask bit '0', the output is inhibited.

13.

Ladder diagram for multiplexing seven segment displays Part 1

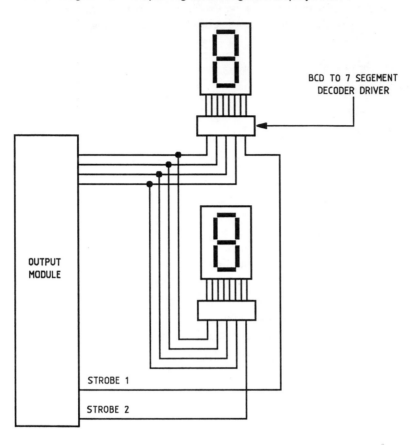

13. Ladder diagram for multiplexing seven segment displays Part 2

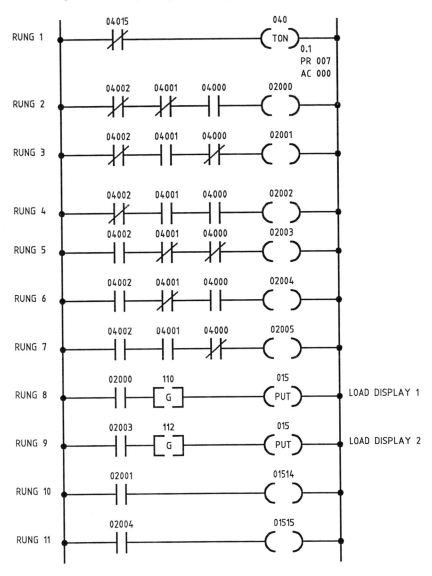

15. A drum sequencer is driven by an electric motor which will cause pegs located on the drum to operate cams causing contacts to close. The absence of a peg causes no output or a zero in this step of the sequence word. A peg causes a contact closure or a one in this step of the sequence word.

Chapter 10

Multiple choice

1. (c) 7 this does not include start, stop or parity
3. (b) 10001
5. (d) frequency division multiplexing

True or false

7. False
9. True

Written response

11. The function of a modem (modulator/demodulator) is to encode digital signals (1s and 0s) into audible tones to allow the digital information to be transferred across a transmission medium. The operation of the modem is as follows:

- Oscillators of different frequency are used to produce the two tones to be sent to line
- A digital switching device is used to switch one oscillator to line if the bit of the character to be sent is a 'mark' or the other oscillator if the bit to be sent is a 'space'. This is the encoding process.
- To decode the signal, one tone, when detected, will cause the digital switching device to produce a mark and the other tone will cause the digital switching device to produce a space.

13.

RS232
Minimum connection

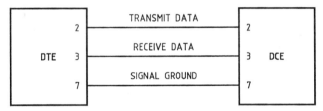

15. The local area network (LAN) can be used over distances of 5 kilometers (3 miles) at data rates of 10 mega bits per second. The LAN is defined by the OSI model. Careful selection of the LAN transmission medium will afford security and very high noise immunity (e.g. optical fibre).

Chapter 11

Multiple choice

1. (a) hardware
   (b) software

3. (a) touching the disk surface
   (b) exposing the disk to a magnetic field

True or false

5. True
7. True

Written response

9. A computer in an industrial environment can have stored data corrupted or its operation disrupted by:

- electromagnetic fields;
- electrostatic fields;
- temperature extremes;
- chemical liquids and gases.

A clean environment should reduce the effect of:

- airborne particles;
- corrosive liquids and gases.

In some cases the clean environment will also be temperature controlled.

Chapter 12

Multiple choice

1. (b) ensure the expected machine position is the same as the actual position
   (b) point to point
3. (a) linear
   (b) parabolic
The third method of interpolation is circular
5. (d) master control unit controlling the robots operations

True or false

7. True
9. False

Written response

11. Improved mass storage. The CNC machine, being microprocessor-based, allows for the storage of programs on floppy disk or magnetic tape which can be retrieved for editing purposes. It also allows those programs to be accessed by other devices via a data communications network.

Ease of editing. As a program can be retrieved, edited, and resaved at the CNC terminal program, alterations become easier.

Axis calibrated. The axis of the CNC machine can be calibrated. This allows repeatable machine motion accuracy.

Reusable machine patterns. A machine pattern once stored can be recalled when required and reused without costly programming time.

Information management. Data communications between CNC devices and other equipment in the manufacturing process can assist in production planning

13. A rotary encoder consists of a slotted disk and a light source. The light passes through the slotted disk onto a light sensitive semiconductor producing a digital output which can be used for servo motor control.

A resolver is a mechanical-to-electrical transducer which produces an analog output with a value which is dependent on the input signal derived from a lead screw.

15. 1. A robot is a device that performs the functions ordinarily ascribed to human beings, or operates with what appears to be almost human intelligence

2. An automatic machine with a certain degree of autonomy, designed for active interaction in the environment.

Chapter 13

Multiple choice

1. (a) technical requirements
   (b) economic impact
   (c) integration into the workplace

The PLC instruction set would be a secondary consideration and would be considered along with other technical requirements of the PLC system

3. (d) conductive dust

5. (a) hardwire

True or false

7. True

9. False

11. False. As many as necessary are permanently wired in series with the PLC system power so the operation of any one will stop the machine or process.

Written response

13. The function of the isolation transformer is to:
- provide physical isolation between the PLC system and power distribution system;
- step up or step down the mains voltage to the voltage required by the PLC;
- provide a constant secondary voltage to reduce the effect of mains power distribution fluctuations.

15. The PLC power supply must be sufficiently large to provide the current to operate the inputs and outputs. To size the power supply it is advisable to base the calculation on the current required when all inputs and outputs are on. It is unlikely that this condition will occur, however, if the power supply is not fully loaded future expansion can be accommodated.

Chapter 14

Multiple choice

1. (a) the user power supply voltage (This will be observed when the input is active or short circuited).
No voltage will be observed while the field contact is open.

3. (a) field wiring

5. (a) AC common or neutral and the output terminal with a 10 load resistor in parallel

True or false

7. False
9. True

Written response

11. Error codes or messages are displayed to assist the PLC user to detect and isolate fault conditions.

13. To troubleshoot a PLC system if a fault occurs:
1. Consider all symptoms – consider the symptoms in terms of the entire PLC system.
2. Make the system safe for troubleshooting.
3. Localize the problem to a rack or unit within the system.
4. Isolate the fault to a unit or module.
5. Carry out corrective action. In most cases this will be achieved by replacing the unit or module. This is done to keep down time to a minimum.
6. Reinstate or recommission the system.
7. Repair the faulty equipment and return to spare parts stock.

15. The program terminal can be used when troubleshooting for:
- program observation.
- forcing 'on' or 'off' data table bits.

Appendix 4

Glossary of terms

AC INPUT ASSEMBLY: A module which converts various AC signals from discrete input devices to logic level for use by the programmable controller processor.

AC OUTPUT ASSEMBLY: A module which converts the logic levels of the programmable controller processor to an output signal to control a discrete output.

ADDRESS: A memory location.

ANALOG INPUT ASSEMBLY: A module which converts an analog input signal its digital equivalent for use by the programmable controller processor.

ANALOG OUTPUT ASSEMBLY: A module which provides an output proportional to a digital value provided to the module.

ANALOG SIGNAL: A continuous voltage or current signal that depends directly on magnitude to represent some condition.

ARITHMETIC FUNCTION: The ability of a PC to do addition, subtraction, multiplication, division and other more complex mathematical functions.

ASCII: Abbreviation for American standard code for information interchange. It is a seven or eight-bit code.

BAUD: A unit of data transmission speed equal to the number of code elements per second.

BINARY: A numbering system that uses a base number of two. There are two digits (1 and 0) in the binary system.

BINARY CODED DECIMAL (BCD): BCD is a system of presenting decimal data in binary code.

BIT: An acronym for binary digit. A bit can assume one of two possible states; ON or OFF; high or low; logic 1 or logic 0.

BOOLEAN ALGEBRA: Shorthand notation for expressing logic functions.

BRANCH: A parallel logic path within a rung of user program.

BUS: Electrical medium used for data transmission and reception.

BYTE: A sequence of binary digits, usually 8, operated upon as a unit.

CAD: Acronym for computer aided design. CAD is sometimes used to describe computer aided drafting.

CAM: Acronym for computer aided management. A system where a number of office and shop floor functions are controlled by a central computer.

CASCADING: A programming technique that extends the ranges of TIMER and COUNTER instructions beyond the maximum values that may be accumulated.

CASSETTE RECORDER: A peripheral device for transferring information between PC memory and magnetic tape. The program data is stored on the magnetic tape in the form of audio tones.

CENTRAL PROCESSING UNIT (CPU): A term for the processor.

CHARACTER: One symbol or a set of symbols, such as a letter of the alphabet or a decimal numeral.

CIM: Acronym for computer integrated manufacture.

CLOCK: A device that generates pulses that are used as periodic signals for synchronization or timing.

CLOCK RATE: The speed at which the microprocessor system operates.

CMOS: An abbreviation for complimentary metal oxide semiconductor.

CNC: Abbreviation for computer numerical control.

CODE: A system of symbols (bits) for representing data (characters).

COMPARE FUNCTION: A program instruction which compares numerical values for 'equal', 'less than', 'greater than', etc.

COMPUTER INTERFACE: A device designed for data communication between a central computer and another unit such as a PC processor.

CORE MEMORY: A type of memory used to store information in represent a logical '1' or '0'. This type of memory is NON-VOLATILE.

COUNTER: A device that can count up or down in response to an input signal and opens or closes a contact when a predetermined count is reached.

CRT: The abbreviation for cathode ray tube, which is an electronic display tube similar to the familiar TV picture tube.

CRT: a terminal containing a cathode ray tube to display programs as ladder diagrams which use instruction symbols similar to relay symbols.

CURSOR: A means for indicating on a CRT screen the point at which data entry or editing will occur.

DATA MANIPULATION: The process of altering and/or exchanging data between STORAGE WORDS or REGISTERS.

DATA TABLE: Manufacturer's allocations of the programmable controller memory.

DATA TRANSFER: The data communicating process of exchanging information between PC memory words or files.

DEBOUNCING: A method of eliminating unwanted multiple closures of mechanical contacts.

DEBUG: The action of searching for and correcting programming errors.

DIGITAL: The representation of numerical quantities by means of discrete numbers.

DISCRETE INPUTS OR OUTPUTS: 'Real world' inputs or outputs that are wired to input and output assemblies.

DUMP: Recording information stored in memory onto magnetic tape disk or some other form of mass storage.

DUPLEX: Two-way data communication.

ELECTRICAL OPTICAL ISOLATOR: A device which couples input to output using a semiconductor light source and detector in the same package.

ENABLE: A circuit that allows a function or operation to be activated.

EVEN PARITY: The condition that occurs when the sum of the number of '1s' in a binary word are always even.

EXAMINE OFF: is a programmable controller instruction is true when the addressed bit is off and false when the addressed bit is on.

EXAMINE ON: is a programmable controller instruction is true when the addressed bit is on and false when the addressed bit is off.

FALSE: As related to programmable controller rungs means power flows through the condition area of the rung to make the rung output active.

FAULT: An incorrect operating characteristic which interferes with normal operation.

FILE: A group of consecutive words that can be accessed under the control of the user program and operated on as a block of data.

FLOPPY DISK: A form of magnetic mass storage.

FORCE: A mode of operation or instruction on the programmer that allows the operator to control the state of contact.

FORCE-OFF FUNCTION: A feature which allows the user to de-energise, independent of the programmable controller program, any input or output by accessing the data table.

FORCE-ON FUNCTION: A feature which allows the user to energize, independent of the PC program, any input or output by accessing the data table.

FMS: An acronym for flexible manufacturing systems which are systems that may contain PLCs which can be adapted to suit the product being made.

FULL DUPLEX: A mode of data communications in which data may be simultaneously transmitted and received.

GROUND: A low resistance path, intentional or accidental, between an electric circuit or equipment chassis and the earth.

HALF DUPLEX: A mode of data transmission which communicates in two directions, but in only one direction at a time.

HARD CONTACTS: Physical switch contacts.

HARD COPY: Paper based printed material produced from a programmable controller.

HARDWARE: The mechanical, electrical and electronic devices which compose a programmable controller system.

HARD-WIRED: Electrical devices interconnected through physical wiring.

HEXADECIMAL: A base 16 numbering system that represents possible four-bit combinations with sixteen digits (0–9 and A–F).

HIGH: A signal type where the higher of two voltages represents a logical '1' (ON).

IEEE: Institute of electrical and electronics engineers.

IMAGE TABLE: An area in programmable controller memory dedicated to input/output data.

286

INPUT DEVICES: Analog and digital. Digital devices such as limit switches, toggle switches, pressure switches, pushbuttons and thumbwheel switches, that supply input information to a programmable controller. Analog inputs from transducers provide a varying voltage or current with respect to time.

INSTRUCTION: A command that will cause a programmable controller to perform a specific operation.

INTERFACING: Interconnecting a programmable controller with its input and output devices.

INTERNAL DEVICES: Counters, timers, etc. within the programmable controller.

I/O: An abbreviation for input/output.

I/O MODULE: The printed circuit assembly that interfaces between the user devices and the programmable controller.

k: Abbreviation for KILO ($\times$ 1000). This abbreviation is used to denote sizes of memory, eg 1k = 1024.

KILO: Measurement to designate quantities 1000 times as great.

LADDER DIAGRAM: A control schematic normally drawn as a series of contacts and coils arranged between two vertical supply lines. The left of the rung may contain condition logic and the output is located at the right.

LADDER DIAGRAM PROGRAMMING: A method of writing a user program in relay ladder diagram format.

LATCH: A device that will continue to store a state after the input signal is removed. The input state is stored until the latch is reset.

LATCH INSTRUCTION: A PC instruction which causes an output to say ON, regardless of how briefly the instruction is enabled and the state of the condition logic in the rung. Latch instruction must be un-latched to turn off the instruction.

LEAST SIGNIFICANT DIGIT (LSD): The digit which represents the smallest value.

LED: Abbreviation for light-emitting diode.

LIMIT SWITCH: A switch which is actuated by some part or motion of a machine or equipment to alter the electrical circuit associated with it.

LCD: Abbreviation for liquid-crystal-display.

LOAD: The output power delivered to a machine or apparatus. A device placed in a circuit or connected to a machine to absorb power. To place information in the form of data into a processor's memory.

LOGIC LEVEL: The voltage magnitude associated with digital signals representing ones and zeros. In positive logic the ones are represented by a positive voltage at or near the supply voltage and the zeros by the absence of voltage termed zero volts.

LOW: a signal type where the lower of two voltages indicates a logic state of '0' (OFF).

MAGNETIC TAPE: Tape made of plastic and coated with magnetic material; used as a form of mass storage of data.

MALFUNCTION: Any incorrect function or operation within electronic, electrical, or mechanical hardware.

MANIPULATION: The process of controlling and monitoring bits, words, or files by means of the user's program.

MASS STORAGE: A means of storing large amounts of data on magnetic tape, floppy disks, etc.

MCR: Abbreviation for master control relay.

MEMORY: Electronic circuit elements which have data storage and retrieval capability.

MICROPROCESSOR: A large scale integrated circuit (LSI) used as the control device in solid state systems including programmable controllers.

MICROSECOND: One millionth of a second.

MILLIAMPERE (mA): One thousandth of an ampere.

MILLISECOND (mS): One thousandth of a second.

MODE: The selected method of operation, eg., RUN, TEST, or PROGRAM.

MODEM: Acronym for modulator/demodulator. A device used to transmit and receive data.

MODULE: An interchangeable 'plug-in' electronic item.

MOST SIGNIFICANT DIGIT (MSD): The digit representing the greatest value.

MULTIPLEXING: A time- or frequency-based system for a number of inputs to share a single channel.

NC: Abbreviation for numerical control.

NEMA STANDARDS: standards for electrical equipment approved by the members of the National Electrical Manufacturers Association (NEMA).

NETWORK: A group of connected elements used to perform a function. A network can consist of a number of elements and outputs. May also apply to a data communications system.

NODE: A common connection point between two or more contacts or elements in a circuit. Also applies to the termination point of data communicating devices connected to a network.

NOISE: Unwanted signals; any disturbance, which causes interference with the desired signal or operation.

NON-VOLATILE MEMORY: a memory that is designed to retain its information while its power supply is turned off.

OCTAL NUMBERING SYSTEM: One which uses a base eight.

ODD PARITY: Condition existing when the sum of the number of '1s' in a binary word is always odd.

OFF-DELAY TIMER: In a programmable controller, an instruction which counts or the delay is started whenever the rung goes FALSE.

OFF-LINE PROGRAMMING: Programming that is done with the processor stopped and unable to produce an output.

ON-DELAY TIMER: In a programmable controller, an instruction which counts or the delay is started whenever the rung goes TRUE.

ON-LINE PROGRAMMING: A method of programming by which rungs in the program may be edited while the processor is running and controlling outputs under program control.

OPTICAL COUPLER: A device that provides electrical isolation by making use of a light source and a light-sensitive semiconductor in the same integrated circuit.

OUTPUT: A signal provided from the controller to the 'real world'.

288

OUTPUT DEVICES: Devices such as solenoids, motor starters, etc; they are the load for the programmable controller.

PARALLEL OPERATION: Type of data communication where 8 bits (1 byte) are transmitted simultaneously.

PARITY: A method of error detecting used in data transmission and receiving.

PC: Abbreviation for PROGRAMMABLE CONTROLLER (PLC). Sometimes used as the abbreviation for personal computer.

PLC: Programmable logic controller, abbreviation for PROGRAMMABLE CONTROLLER

POINTER: A register address at which there is contents location information.

PORT: An input/output connection on a PLC system.

PRIORITY: Order of importance.

PROCESSOR: The part of the programmable controller that performs logic solving, program storage, and supervisory functions.

PROGRAM: A sequence of instructions to be executed by the processor to control a machine or process.

PROGRAM PANEL (PROGRAMMER): A device for inserting, monitoring, and editing a program in a programmable controller.

PROGRAM SCAN TIME: The time for the processor to execute all instructions in the program once.

PROGRAMMABLE CONTROLLER: A solid state control system, designed to be used in the industrial environment, which has a user programmable memory for storage of instructions to implement specific functions such as I/O control logic, timing, counting, arithmetic, etc. The programmable controller is used to control a machine or process.

PROM: Acronym for programmable read only memory.

PROTOCOL: A defined means of establishing criteria for receiving and transmitting data through communication channels.

RS–232C: An Electronic Industries Association (EIA) standard for data transfer and communication.

RACK: A programmable controller chassis that contains modules.

RAM: Acronym for random access memory.

READ/WRITE MEMORY: A memory in which data can be placed (write mode) or accessed (read mode).

REGISTER: A location within the controller allocated to the storage of digital information.

REPORT: A display or printout containing information in a user-designed format.

RETENTIVE OUTPUT: An output that remains in its last state (ON or OFF).

RETENTIVE TIMER: A programmable controller instruction which continues to time while the rung is true and retains the accumulated value while the rung is false.

ROM: Acronym for READ ONLY MEMORY.

RUNG: A grouping of programmable controller instructions which control an output or storage bit. A rung may be unconditional whereby the output will not be preceded by condition logic.

SCAN: The scanning operation, as performed by the processor, is the sequential examination of user program stored in memory and the status of inputs and outputs to produce the desired function.

SCAN TIME: The time required to make one complete scan through user program and to update all outputs based on the condition of the inputs together with the user program function.

SCHEMATIC: A diagram of a circuit using symbols to illustrate circuit components.

SCR: Silicon controlled rectifier used to switch power in AC and DC circuits.

SEQUENCER: A controller which operates an application through a fixed sequence of events.

SERIAL OPERATION: Type of data transfer within a programmable controller whereby the bits are handled one after the other in a serial train.

SHIELDING: The practice of confining the electrical field around a conductor or stopping the intrusion of external electrical fields by putting a conducting layer over the cable insulation.

SIMPLEX: A mode of data communications but in one direction at a time.

SOFTWARE: The user program which controls the operation of a programmable controller.

SOLID STATE: Circuitry designed using only integrated circuits, transistor, diodes, etc.

STATE: The logic '1' or '0' condition in programmable controller memory.

SWITCHING: The act of turning ON and OFF a device.

THUMBWHEEL SWITCH: A rotating numeric switch used to input digital information to a programmable controller.

TIMER: In a programmable controller, a timer is user-programmed and is internal to the processor.

TOGGLE SWITCH: A panel-mounted switch with an extended lever; normally used for ON/OFF switching.

TRANSDUCER: An energy conversion device.

TRANSFORMER ISOLATION: A method of isolating I/O devices from the controller.

TRIAC: A solid state component capable of switching alternating current.

TRUE: As related to programmable controller instructions, the instruction is active or enabled. In relation to a rung the rung is considered to pass power.

TRUTH TABLE: A method of representing logic functions by listing all possible combinations of inputs and resultant outputs.

TTL: Abbreviation for transistor transistor logic. A family of integrated circuit logic, usually 5 volts is high or '1' and 0 volts is low or '0'.

UNLATCH INSTRUCTION: A programmable controller instruction which causes a latched output to release.

VOLATILE MEMORY: A memory that loses its information if the power is removed from it.

WORD: A grouping or a number of bits in a sequence that is treated as a unit.

WRITE: Placing data into a register or memory location for storage.

Index